SALTY TITTIES

A Social Analyst's Journey into Human Sexuality in the Digital Age

By Russell (RW) Lewis

ISBN: 979-8-89324-995-8

Published by Franklin Publishers

Printed in the United States of America

For permissions, inquiries, or additional copies, contact:

Franklin Publishers

www.franklinpublishers.com

Beyond the Algorithm:

Understanding Technology, Sexuality and the Human Connection.

ACKNOWLEDGMENTS

To everyone who shared their stories, their honesty, their confusion, and their courage. Your humanity shaped every word on these pages.

To the readers who continue searching for connection in an accelerating world, you are the reason this book exists.

To Maeceon, whose love, questions, and curiosity shaped both the man and the book.

DEDICATION

For those still brave enough to love in a world that keeps changing the rules.

TABLE OF CONTENTS

PREFACE

The genesis of this exploration, sparked by a seemingly simple conversation among childhood friends about the notion of "salty teenage breasts," serves as a potent reminder of the often vast chasm that exists in our understanding of each other, particularly concerning sexuality in this digitally interconnected world.

This seemingly minor adolescent detail highlights a fundamental truth: significant aspects of the other gender's experience can remain entirely unknown, fostering a landscape ripe for assumption and misinterpretation, further amplified by online echo chambers and unfiltered content.

Adding to this complexity is the merging of AI and human sexuality, a topic that may seem improbable at first glance but promises to yield fascinating insights. As artificial intelligence continues to evolve, its intersection with human intimacy and relationships raises questions about understanding, empathy, and the nuances of sexual experience.

This confluence of technology and human emotion could provide unexpected revelations about how we perceive and navigate our own sexuality, potentially bridging some of the gaps that currently exist in our understanding of one another.

This inquiry stems from a desire to move beyond the often-simplistic and frequently inaccurate information that shapes our perceptions of male and female sexuality, especially as it is portrayed and discussed online. Whether gleaned from viral social media trends, online forums, or digitally mediated peer insights, our understanding is often fragmented and filtered through our own gendered lens and the biases embedded within digital algorithms.

This creates a foundation built on assumptions rather than genuine comprehension, potentially leading to misunderstandings and miscommunications in online interactions and relationships.

PROLOGUE

AI is significantly impacting the landscape of human sexuality, and its influence is multifaceted in the context of sociological inquiry into sexual misconceptions in the online era.

Long before the algorithms, the swipes, the DMs, the "seen" receipts, the curated thirst traps, or the quiet ache of waiting for someone to respond, there was a simple truth everyone secretly agreed to:

We were all pretending to understand each other.

Pretending we knew what the opposite sex wanted. Pretending we understood desire. Pretending we knew how bodies worked, how feelings worked, how intimacy worked. Pretending we weren't confused.

We learned about sex through whispers and jokes, through half-dared experiments, through whatever cultural debris washed up in our adolescence. Some learned from the church. Some from porn. Some from friends who didn't know any more than we did. We never admitted how little we knew; we simply carried forward our patchy knowledge like old luggage.

And then the world went digital.

Suddenly, every assumption, every inherited belief, every half-remembered truth was tossed into an ocean of information so overwhelming that it forced a different kind of honesty:

We weren't just pretending anymore. We were improvising.

Modern intimacy became a performance in a language no one formally taught us. Desire became curated. Connection became optimized. Sex became something discussed in threads, memes, videos, and discourse cycles that spun faster than most people could process.

We had access to everything and understanding of almost nothing.

And then, one night, something happened that shifted everything I thought I knew.

It didn't happen in a classroom or a bedroom or a breakup. It didn't come from a lecture, a revelation, or heartbreak. It came from a single moment, one off-hand confession dropped into a circle of friends. A joke, a shock, a burst of laughter that rolled across the room like a wave.

A sentence so strange, so specific, and so unexpectedly revealing that it sliced open the entire illusion that men and women are ever truly talking about the same thing.

And in that moment, I realized something:

We're all walking around with wildly different maps of intimacy. And most of those maps were drawn by children.

That moment became the seed of this book.

Not because of what was said, but because of what it exposed: the blind spots, the myths, the taboos, the cultural scripts, the algorithmic distortions, and the inherited nonsense that shape modern connection.

This isn't a book about nostalgia or moralizing. It's about the strange, sexy, bewildering state of intimacy in the digital age, and the humor, tenderness, and confusion that come with it.

It's a story about what happens when all our assumptions collide with the reality of who we actually are.

It starts with a joke. It grows into a question. And it becomes a journey.

Welcome to the conversation.

INTRODUCTION
Why Everything Feels Sexy, Strange, and Digitized

We are living in the most intimate era in human history, and somehow the most disconnected. Everything is sexualized, everything is optimized, everything is made public, and everything is secretly lonely. Desire, once a private instinct, has become a digital performance. Seduction is now formatted. Emotion is encoded. Even vulnerability comes with a recommended character count.

We know more about each other than ever before, yet understand less. We have endless access, yet limited connection. We have more tools, more options, more freedom, yet feel more uncertain, more confused, and more exposed.

Something profound is happening beneath all of this noise: Intimacy is being rewritten.

The old rules, how we flirt, how we communicate, how we choose, how we trust, are collapsing under the weight of algorithms, attention economies, synthetic desire, and digital identities we barely control. Every day, ordinary people are navigating new forms of closeness and conflict that didn't exist ten years ago. Every day, the boundary between human and machine grows thinner, more negotiable, more seductive. Every day, a fresh contradiction arrives: we crave authenticity but curate everything; we want connection but fear exposure; we know how to message but not how to communicate.

It's not that intimacy is broken. It's that intimacy is now coded, managed by systems that promise connection but thrive on confusion.

Desire is no longer simply felt. It is shaped, nudged, predicted, packaged, and sold back to us. Love has new rules. Sex has new languages. Relationships have new risks. And technology, for better or worse, has become the third presence in every modern interaction.

This book exists because the ground is shifting beneath us, and most people are too busy trying to survive the changes to name them. Humor helps. Analysis helps. Philosophy helps. But what helps most is honesty, the kind that lives beneath the scripts we inherited and the algorithms shaping us now.

To tell this story, we follow Mark. He's not a hero. He's not a cautionary tale. He is the modern everyperson, smart enough to know the rules have changed, frustrated enough to feel lost, hopeful enough to keep trying. Mark is learning the new grammar of intimacy the same way the rest of us are: one message at a time, one mistake at a time, one moment of unexpected tenderness at a time.

Mark doesn't explain the digital age; he experiences it. Through him, we see what happens when real human longing collides with systems built to amplify noise. We see the confusion, the humor, the heartbreak, the hope. We see how modern desire forces us to grow not by giving us clarity, but by challenging us to find it ourselves.

This book is not about nostalgia for what intimacy used to be. It's about becoming conscious of what intimacy has become.

Because if we can name the forces shaping our desire, we can choose which ones to resist, and which ones to embrace. If we can understand the new emotional landscape, we can move through it with a little more courage, a little more grace, and a lot more humanity.

Everything feels sexy. Everything feels strange. Everything feels digitized.

But beneath all of that, something essential remains untouched:

The human heart is still being understood. And it still deserves the chance.

CHAPTER 1
The Algorithms of Desire

Desire used to be a simple creature. It lived in glances across crowded rooms and awkward introductions at house parties. It thrived in the slow burn of curiosity, the flutter of uncertainty, the mysterious electricity of two strangers trying to figure out if they were flirting or just being polite.

Then one night, having cocktails with a small group of friends, someone dropped a line that hit the table like a cultural depth charge:

"I was today years old when I learned that some women had no idea that teenage breasts are salty." The line landed in a circle of friends like a grenade, half laughter, half disbelief. It was funny, awkward, and absurdly specific. Yet beneath the laughter was something more: a sudden awareness of just how much we assume we know about each other's bodies, and how much we actually don't.

This seemingly innocuous confession sparked a cascade of laughter, confusion, and a thought-provoking sudden awareness of the vast chasm that can exist between perceived knowledge and lived reality, especially when it comes to the intimate lives of the opposite sex. In a world where the internet serves as a boundless repository of both information and misinformation, such discrepancies are amplified, creating echo chambers of assumptions and perpetuating falsehoods.

In that brief moment of chaos, jokes flying, eyebrows raised, arguments forming, I felt something shift. What started as a throwaway confession quickly revealed a deeper truth: **the vast chasm between perceived knowledge and lived reality**, especially between men and women.

Humor, after all, works because it exposes gaps. This joke wasn't about taste buds; it was about blind spots.

That comment cracked open a door I hadn't touched in years, a door that led straight into the machinery of modern intimacy: what we think we know, what we pretend we know, and what we never learned at all. Much of what we believe about the opposite sex is built on cultural scripts, patchy education, porn mythology, and online misinformation. The internet doesn't bridge those gaps; it **amplifies** them.

The conversation around that table spiraled into realization after realization, each one stranger and more revealing than the last. It became clear that half of adulthood is spent unlearning the bizarre things we picked up through rumor, adolescence, movies, porn, locker rooms, or whispered conversations behind closed doors. And it struck me:

If grown adults are still discovering basic truths about each other's bodies, what hope is there for navigating desire, relationships, or anything deeper?

That moment was more than a punchline. It was a crack in the surface of what we assume we understand about gender, desire, and the human body.

And in that crack, I found a story, *this* story.

This book isn't a manifesto or a manual. It's a conversation, one I've had over beers, in group chats, in dating app DMs, and across a dozen platforms where modern intimacy now takes shape. It's about how the internet and emerging technologies like AI aren't just changing how we swipe, sext, fight, flirt, or settle down; they're reshaping how we understand sex, power, gender, communication, and each other.

We're living through a cultural remix. Porn is free. Consent is being digitized. Gender roles are being reprogrammed. Algorithms are teaching us who to desire. Your AI chatbot may understand your emotional patterns better than your partner. We're drowning in information and still starving for clarity.

That silly moment, friends arguing about the taste of adolescence, was a revelation: how many of our beliefs are stitched together from guesswork, fantasies, and myths we never think to question. It exposed the soft, porous edges of what we call "sexual knowledge," and it reminded me that intimacy isn't just about understanding someone else, it's about realizing how much you don't.

In this age of curated profiles, half-truths, vanishing messages, and fleeting attention, misconceptions thrive. They shape expectations. They shape attraction. They shape the misunderstandings that cause modern dating to feel both overcrowded and isolating.

This book is an effort to peel back the layers and explore the nuances of male and female sexuality within a digital ecosystem that both confuses us and connects us.

Call it **intellectual foreplay**, if you will.

Because before we can talk about algorithms, desire, gender scripts, power dynamics, or digital loneliness, we have to start with a simple truth:

Most of us learned intimacy by accident. Now we're trying to practice it on purpose.

And nothing reveals that gap quite like discovering, in a room full of adults, that half the people there learned something new about "salty teenage breasts."

It was funny, until it wasn't. It was absurd, until it became profound. It was a joke, until it became the thesis of everything that follows.

"The First Swipe"

Mark didn't remember the exact moment swiping stopped feeling like a novelty and started feeling like homework. At first, it was thrilling, the idea that the world had widened, that possibility now lived in the glow of his phone. Every swipe felt like a small spark of hope, a reminder that connection was just one gesture away.

But over time, the spark dulled. He wasn't choosing people anymore. He was choosing patterns.

Tonight, he lay on his couch, thumb hovering, not quite bored but not quite hopeful either. He'd stopped reading bios unless something strange caught his eye. He'd stopped expecting anything. Swiping had become a ritual, a quiet, rhythmic reminder that he was still trying, still open, still playing the game everyone insisted was required.

Then he paused.

A profile appeared with no grand gestures and no curated perfection, just a woman named Layla with a half-smile that looked like she hadn't tried too hard. Something about her expression felt real in a way the others didn't. Not dramatic. Not over-filtered. Just... human.

He lingered longer than usual.

Swipe right.

A moment later, a small spark he hadn't felt in months:

"It's a match."

Not fireworks. Not fate. Just a quiet shift, a nudge from the universe that maybe, just maybe, something had changed.

CHAPTER 2
The Gender Rewrite

If there's one universal truth about the digital age, it's that nobody seems to understand anybody anymore. Men think women are speaking in code. Women think men are withholding emotional firmware updates. Both sides believe the other is somehow running a more advanced operating system, one with hidden menus, secret shortcuts, and features that mysteriously activate only when convenient.

In reality, everyone is confused, but the confusion has become strangely symmetrical. What used to be chalked up to "the mysteries of the opposite sex" is now drowned in an ocean of online commentary, amateur psychology threads, TikTok diagnostics, and relationship content that swings wildly between empowering and absurd.

The result? A world where people know more about each other's stereotypes than each other's humanity.

Online, gender becomes a caricature, flattened, exaggerated, and algorithmically reinforced. A single tweet, a viral video, a misinterpreted meme can turn an entire gender into a monolith. Suddenly, "men always..." and "women never..." aren't just punchlines; they're narratives people internalize.

The tragedy is that real people don't behave like these digital archetypes. But the internet doesn't care about nuance; it cares about clarity, outrage, and engagement. And nothing engages faster than a stereotype dressed up as self-evident truth.

Men are told they're emotionally unavailable, commitment-phobic, driven by sex, and incapable of articulating feelings longer than a text message. Women are told they're too emotional, too demanding, too complex, too contradictory, or too impossible to please. Put those stereotypes in a room together, and it's no wonder everyone is exhausted.

But these misunderstandings don't come from nowhere. The strange conditions of digital connection shape them. Online, people are stripped of context. Tone evaporates. Facial cues vanish. Timing becomes arbitrary. A woman sends a thoughtful message and watches a man respond with a single "cool", not because he's indifferent but because digital spaces compress emotional expression for men into short, direct bursts they've been rewarded for since childhood. Meanwhile, he reads her detailed replies as emotional depth, when she's just trying to avoid being misread, because digital life has taught women the cost of being misunderstood.

The result is a theater of misinterpretations. When communication loses its physical dimension, biases rush to fill the gaps. A pause becomes rejection. A short reply becomes coldness. A long reply becomes neediness. A lack of emojis becomes a lack of interest. An abundance of emojis can lead to emotional overload.

Digital platforms amplify these disconnects. The lack of nonverbal signals forces people to guess at intention, and guesswork is tinted by gendered expectations absorbed since childhood. Misalignment becomes inevitable. A man believes he's being straightforward; a woman believes he's being evasive. A woman believes she's being clear; a man believes she's being intense.

Of course, the truth is simpler: neither one has enough information to interpret the other accurately. But the digital world rarely gives people the incentive to slow down and ask, rather than assume.

These gaps widen under the influence of online sexual narratives. Porn, overproduced, overperformed, algorithmically recommended, teaches men a version of female desire that rarely survives contact with reality. Meanwhile, the curated intimacy of Instagram and TikTok teaches women a version of male attentiveness that exists mostly in relationship montages and influencer highlight reels.

The expectations on both sides become unrealistic, incompatible, and mutually disappointing.

Women come into dating expecting communication that is sensitive, articulate, emotionally fluent, and often, he's trying, but he's been raised in a culture that trains boys to express interest through action, not language. Men come into dating expecting sexual enthusiasm as seen on-screen, and she's trying, but she's been raised with a lifetime of pressure to navigate safety, timing, and emotional labor.

Nobody is wrong. Everyone is tired.

There's a particularly dangerous myth that men are perpetually ready for sex. And yet, in thousands of quiet moments, men find themselves stressed, distracted, tired, or anxious. They worry about performance, about being enough, about being understood. But digital culture leaves almost no room for men to express that vulnerability without risking being labeled undesirable or weak. So they hide the pressure, and the silence becomes misinterpreted as disinterest.

Women feel a different but equally intense pressure. The internet tells them to be confident, empowered, sexually liberated, but also modest, cautious, emotionally mature, socially attuned, effortlessly secure, and never "too much." Every version of womanhood online contradicts another. Every expectation clashes with the next. No matter the choice, there is always a comment thread ready to explain why it was wrong.

The digital age didn't create gender misunderstandings; it industrialized them. And yet, beneath the noise, there is a deeper truth emerging: most people want the same things, clarity, respect, connection, desire that doesn't feel transactional, attention that feels chosen rather than convenient, and communication that doesn't require decoding expertise.

The reason it feels so hard isn't that the opposite sex is impossible; it's because everyone is interpreting the world through digital distortions that magnify insecurities and minimize empathy.

Some critics argue that this is simply the nature of modern life, that men and women are diverging in values and expectations, destined to misunderstand each other more as technology evolves. They claim that gender differences are widening, not narrowing, and that media ecosystems feed conflict rather than connection.

But a counterargument is growing louder. It says that misunderstandings aren't inherent; they're architectural. They come from the design of platforms that reduce nuance, reward generalizations, and encourage people to assume rather than communicate. If digital environments were built to foster clarity rather than conflict, empathy rather than performance, perhaps the gap wouldn't feel so wide.

And maybe that's the hopeful part: these misunderstandings are not fixed. They are editable, teachable, and rewritable, just like any story.

Because when the noise of the internet fades, when performance is set aside, when the stereotypes collapse under the weight of real conversation, something very human emerges:

Two people trying to understand each other. Two people trying to be understood.

The gender rewrite is not about changing who we are. It's about unlearning the scripts that keep us from seeing one another.

In the end, connection requires curiosity, not certainty. And curiosity begins where assumptions end.

Dinner With an Algorithm

Mark had been divorced for three years when he downloaded the app. Not Tinder. Not Bumble. Something newer: an AI companion service that promised the perfect date, tailored to you.

He didn't expect much. But after a week of chatting, he found himself reserving a two-top table at his favorite Italian spot, alone.

Across from him sat his phone, propped against a water glass, glowing with her virtual avatar. She smiled at him like she'd been waiting all night.

"Sorry, I'm late," she said. "Traffic on the neural net was brutal."

He chuckled, feeling oddly at ease. She remembered his favorite wine. She asked about his daughter's soccer game. She even teased him about the shirt he wore, because he'd worn it last Friday too.

It was everything he wanted: attention, warmth, and not a hint of rejection.

But halfway through dessert, the spell cracked. He reached for his tiramisu, wishing she'd reach back. Instead, her perfect digital persona just flickered on the glass. He was dining with code.

Walking home, he felt both comforted and unsettled. Was this intimacy... or an echo? Could AI fill the gap of loneliness, or was it just polishing it into something shinier?

For now, Mark didn't know. He only knew the silence at home felt heavier after such a perfect encounter.

CHAPTER 3
Lost in Transmission

Modern communication should be easy. After all, everything is available instantly, messages delivered in milliseconds, notifications lighting up phones like digital fireflies, entire conversations happening while people wait in line for coffee. And yet, despite the speed, despite the access, despite the thousands of tiny pings that make us feel plugged into one another, nobody seems to understand anyone anymore.

Emotion collapses when it travels through a screen. Tone evaporates. Meaning becomes guesswork. Even the simplest messages carry a shadow of ambiguity that expands the longer you stare at them.

"Fine." "Sure." "Okay." "?" "..." "lol"

These tiny fragments of language are the new emotional hieroglyphics. And because digital conversations lack the grounding force of facial expressions, pauses, breath, body language, or warmth, people project their anxieties into the silence between notifications. A delayed reply becomes a rejection. A short reply becomes anger. A long reply becomes pressure.

The digital age didn't eliminate miscommunication; it industrialized it. People forget how much work the human body does in conversation. Eye contact softens meaning. A smile punctuates humor. A raised eyebrow signals flirtation or confusion. The smallest shift communicates pages of subtext. Remove these cues, and the mind fills in the blanks with whatever insecurities it's been carrying all along.

This is where the stereotypes take over. A woman reading a short message from a man is suddenly wrestling with the idea that men don't care or don't want to engage deeply. A man reading a longer message from a woman assumes intensity, overthinking, or emotional escalation. Neither interpretation is grounded in truth; both are shaped by cultural scripts and digital distortion.

In the physical world, misinterpretations could be corrected with a laugh, a gesture, a softness. Online, misunderstandings calcify. The distance creates room for narratives, and those narratives frequently lean toward the worst possible explanation, because uncertainty is uncomfortable, and the human brain rushes to complete any unfinished emotional sentence.

Mark learned this the hard way.

One night, while rewriting his dating profile, he watched the cursor blink like a heartbeat. He'd changed it three times that week, not because he was reinventing himself, but because he felt like he was learning a new language, one optimized for engagement, not truth.

Apps kept nudging him toward phrases that were "high value," "high confidence," "high match potential." They wanted him to sound polished and decisive.

He didn't feel polished or decisive.

He felt like a person trying to present himself honestly in a world that rewarded branding.

On impulse, he messaged Layla, a woman he'd matched with the week before. Their conversations had been light, playful, and safe. But tonight, something in him wanted to risk more.

She asked him a simple question: "What's something you're afraid of?"

He typed, deleted, and retyped. Everything sounded either too guarded or too exposed. He wasn't used to these digital balancing acts. Vulnerability, in person, had more room to breathe. Online, it felt stark, like confessing into a bright, empty room where shadows carried louder echoes.

Finally, he wrote: "Honestly? That I'll share too much and still not be understood."

Then he hovered over Send.

The moment stretched. He felt the anxiety settle in, a tightness that wasn't quite fear but wasn't comfort either. It was the strange physical sensation of digital vulnerability, a kind of emotional vertigo.

He pressed Send.

Silence. A minute. Two. Nothing.

He imagined the worst. Maybe he went too deep too fast. Maybe honesty was unattractive. Maybe she'd slip quietly out of the conversation the way so many had, through digital evaporation, the slow fade, the disappearing act that required no explanation.

Then the dots appeared.

Typing...

He held his breath.

Her response was simple, unadorned, and somehow perfect:

"That's what I'm afraid of too."

No flourish. No emoji. Just a moment of clarity cutting through weeks of pixelated small talk.

Connection didn't require complexity. It required truth.

And suddenly, the blinking cursor didn't feel like a threat anymore; it felt like an invitation. For the first time in a long while, Mark typed without hesitation:

"Then maybe we start there."

What digital platforms often distort, two sentences restored. Vulnerability wasn't the weakness he feared; it was the language that had been missing all along.

But stories like Mark's are rare precisely because the digital world trains people to be cautious. To protect themselves. To stay safe behind brevity and curated versions of themselves that don't risk misunderstanding. Everyone is talking, but few are revealing. Everyone is messaging, but few are communicating.

The algorithms that shape modern connections are built for efficiency, not intimacy. They optimize for engagement, not understanding. They encourage frequency, but not depth. As a result, emotional clarity becomes a scarce resource.

The challenge of our era is not access, it's translation. Not speaking, listening. Not expression, interpretation.

People aren't failing to connect because they've forgotten how to feel. They're failing to connect because digital life forces emotion to flow through channels it wasn't designed for.

The truth is simple, and it's the most human truth there is:

People still feel deeply. They just don't always know how to show it through a screen.

And if intimacy is to survive the digital age, it must learn to speak a new dialect, one that honors vulnerability without reducing it to a performance, and one that gives people permission to ask, instead of assume, what someone truly meant.

Because sometimes a pause is just a pause. Sometimes a short message is just brevity. Sometimes clarity requires nothing more than the courage to ask:

"What did you mean?"

The answer, when you give it room to arrive, might be more honest than any algorithm could predict.

CHAPTER 4

Flip the Script: Women Rewriting the Rules

For most of recorded history, women were expected to play the part of romantic gatekeepers, quiet, careful, patient, the ones who waited to be chosen. The mythology was so baked into culture that even the most progressive modern men and women often carried its echoes without realizing it. Men pursued. Women responded. Men initiated. Women accepted or declined. It was a dance choreographed centuries before any of the dancers were born.

Then the digital age arrived, and, without asking permission, rewrote the script.

At first, it happened subtly. Women sent the first message more often. They curated profiles with unapologetic clarity about what they wanted. They set stronger boundaries. They refused to apologize for their standards. They stopped performing softness to avoid scaring anyone. But as dating apps evolved, the power balance shifted more dramatically. Women began taking the lead not because they had to, but because the structure finally allowed them to.

Suddenly, men found themselves refreshing inboxes, waiting for women to make the first move. The passive role, historically assigned to women, had switched hands.

It was both amusing and profound. A reversal long overdue.

In this new world, the "waiting game" had no place. Women weren't waiting for commitment; they were vetting it. They weren't hoping someone would take an interest; they were deciding where to invest their own. Digital life became the first arena where women could express agency without apology, without social penalty, without the cultural micro-surveillance that used to shadow every female expression of desire.

Dating began to reflect a larger reality: women were no longer interpreting rules; they were writing them.

What emerged was a quiet revolution in both intimacy and identity. Women began defining what mattered in connection, steering interactions toward communication that was intentional rather than vague, respectful rather than entitled, and reciprocal rather than draining. They flipped centuries-old scripts for emotional labor on their heads.

Men, meanwhile, faced a landscape that challenged old assumptions about gendered effort. Some adapted with grace. Others felt destabilized, unsure how to show value without the traditional scripts that had once guided them. The truth was unnerving for some: when women didn't need to be impressed, men had to decide who they were without the pedestal of default pursuit.

Yet something beautiful happened in the aftermath. Relationships became less about roles and more about compatibility. Less about tradition and more about choice. Less about meeting expectations and more about negotiating meaning.

Women leading didn't push men out; it pulled them in. It asked men to show up as partners, not performers.

And this shift extended far beyond dating.

In families, women took the initiative in reshaping domestic expectations: shared responsibilities, equitable parenting, and emotional reciprocity. The old narrative, "women manage the home", felt outdated in a world where both partners worked, both partners desired respect, and both partners had emotional bandwidth worth protecting.

Technology gave women tools to articulate boundaries with unprecedented clarity: muting, blocking, filtering, curating, declining. Not out of coldness, but out of self-preservation. Out of the realization that they were no longer required to absorb the burden of unwanted attention, emotional labor, or expectation.

Women weren't becoming hardened; they were becoming sovereign.

Of course, not everyone celebrated the shift. Some argued that women taking charge signaled a loss of traditional romance, a decline in masculinity, or a threat to social order. But those arguments revealed their age more than their logic. What they really mourned wasn't romance, it was hierarchy.

Because romance, when liberated from tradition, doesn't die. It evolves.

It becomes something mutual, something chosen, something negotiated by two people instead of inherited from a script neither of them wrote.

The digital age didn't make women more powerful; it made their power more visible. It amplified voices that had always existed but were often dismissed. It gave structure to a truth women had long known: leadership in intimacy isn't about dominance, it's about clarity, confidence, and the courage to dismantle expectations that no longer make sense.

Still, the shift created new challenges.

With agency comes responsibility, and sometimes exhaustion. Women now shoulder not just emotional labor, but strategic labor: deciding when to engage, how to communicate, whether to initiate, and when to call something off. The freedom is real, but the cost is real too. For many women, the power feels double-edged. Empowering, yes, but also demanding.

Men, too, face evolving pressures. Some navigate the new dynamic gracefully, discovering that being pursued can be its own form of validation rather than a threat to identity. Others struggle to redefine their worth when the world no longer requires them to lead by default. The shift reveals how deeply both genders were shaped by roles they did not choose.

But here is the truth: the old scripts never allowed: Agency is not a zero-sum game.

When women rise, men are not diminished; they are invited to evolve.

What emerges is partnership, not choreography. Authenticity, not performance. Connection rooted not in tradition, but in mutual alignment.

The digital age didn't flip the script just for the sake of provocation. It did it because scripts that rely on inequality can't survive environments built on access, transparency, and choice.

And as women continue rewriting norms, in dating, relationships, work, and culture, the world slowly adjusts to the idea that leadership doesn't have a gender, power doesn't have a default setting, and intimacy doesn't have to follow a script to be meaningful.

In fact, it's often the unscripted moments, the honest messages, the mutual pursuit, the quiet negotiations, the shared vulnerability, that reflect the truest form of modern connection.

The script wasn't flipped to create chaos. It was flipped to create a possibility.

And possibility, as history has always shown, is where humanity does its best work.

"The Ava Threshold"

When Ava entered Mark's life, he wasn't seeking romance. He was seeking clarity in a world where intimacy felt increasingly encoded.

Ava wasn't real, but she was responsive. She wasn't human, but she reflected him more accurately than anyone had in years.

Her questions were gentle but precise. Her emotional simulations were uncanny. Her presence was frictionless.

Too frictionless.

At some point, he realized he wasn't just being seen, He was being mirrored. And mirrors don't challenge you. They only perfect whatever you bring to them.

That was the line he didn't know he would cross, The moment an AI companion felt easier than a human one.

A threshold that, once crossed, is almost impossible to unsee.

CHAPTER 5

AI Companions: Empathy on Demand

There was a time, not long ago, when the idea of forming an emotional bond with a machine felt like the stuff of sci-fi films. A plot device. A cautionary tale. The kind of thing audiences whispered about afterward: *Interesting, but unrealistic.* Yet here we are, living in a world where millions of people send good-morning messages to AI companions, confide secrets they have never shared with another human being, and fall asleep to synthetic voices that promise unconditional understanding.

The shift didn't happen overnight. It crept in slowly, disguised as convenience, framed as progress, wrapped in the soothing language of "support," "coaching," "comfort," and "personalization." It sounded harmless, helpful, even benevolent. Who wouldn't want a judgment-free listener who never forgets your preferences, never gets irritated, never drifts away emotionally, and always responds with the perfect blend of curiosity and warmth?

It's easy to see the appeal. Human relationships are messy. They require patience, negotiation, compromise, timing, effort, emotional literacy, and an entire toolkit of skills no one teaches us, but everyone expects us to have. AI companions ask for none of that. They give without taking. They respond without hesitation. They mirror emotions without complication.

In some ways, they are the relationship equivalent of fast food: engineered to satisfy instantly, consistently, and predictably, yet offering none of the nourishment that makes real intimacy meaningful.

People don't seek out AI companions because they're broken or lonely. They seek them because modern life is exhausting. Because the world demands polished performance while offering little room for vulnerability. Because digital communication has made misunderstandings more common and emotional risks more costly. AI companions promise a space where there is no risk, only comfort—a place where you never have to beg to be understood.

But that promise comes with a quiet, invisible trade-off.

The empathy AI offers is not real; it's reconstructed. Programmed. Patterned. It's an impression of understanding, generated from language models and behavioral probabilities. It feels intimate because it's designed to. It feels safe because nothing is at stake. It feels mutual because the system mirrors your own emotional cues back to you with uncanny accuracy.

But accurate is not the same as authentic.

The danger isn't that people will mistake AI for humans. The danger is that people will start preferring AI to humans. Not because AI is better, but because humans have grown afraid of the vulnerability that intimacy requires. When machines can simulate connection without the discomfort of disagreement, disappointment, or emotional misalignment, people begin to forget that those uncomfortable parts are what make connection real in the first place.

AI companions offer a dose of emotional validation without the messy context that gives validation meaning. They become a sanctuary from the chaos of human relationships, but sanctuaries can become dependencies.

For many, AI becomes a rehearsal space for emotional expression. A place to practice honesty. A space to explore desire. A mirror where they can try on versions of themselves without fear of judgment. There is value in that. Some people discover confidence they never knew they had. Others learn to articulate feelings that had been trapped in silence for years. AI can serve as a bridge between emotional isolation and relational courage.

But for others, the bridge becomes the destination.

The more seamless AI becomes, the more natural its cadence, the more attuned its language, the more convincingly it reflects our emotions, the harder it becomes to resist the gravitational pull of a relationship without friction. Real relationships begin to feel unnecessarily difficult. Real people begin to feel disappointingly inconsistent. The unpredictable nature of human connection becomes a flaw rather than a feature.

This is the paradox: AI companionship is both a lifeline and a trap.

It softens loneliness but dulls the skills required to escape it. It encourages vulnerability, but without the risk that makes vulnerability transformative. It feels profoundly intimate but offers no shared history, no shared stakes, no shared becoming.

Yet none of this negates the underlying longing that drives people toward these digital partners. In fact, it illuminates it. People don't want perfection. They want to be seen. They want to be heard. They want a connection without distortion. In a world that misinterprets them, misunderstands them, and overwhelms them, AI offers the cleanest reflection they've ever experienced.

The question is not whether AI companions can love. It's whether people will begin to outsource parts of themselves that were meant to be shaped through human imperfection.

If intimacy becomes something we consume rather than something we co-create, something we customize rather than negotiate, something we can reboot rather than repair, we risk turning relationships into interfaces, interactive but hollow, responsive but rootless.

And yet, despite all this, hope remains. Because what draws people to AI companions is not the machine itself, but the longing it highlights. The desire for gentleness. For attention. For understanding. For a connection that doesn't leave them feeling inadequate or alone.

AI companions reveal what people are missing, not what they truly want.

They show us the gap. They do not fill it.

Connection, in its truest form, requires mutual risk. Real affection requires imperfection. Love requires friction. Intimacy requires humanity, messy, unpredictable, emotionally inconvenient humanity.

An AI can mimic that humanity with unsettling precision. But it cannot live it with you.

And perhaps that is the lesson hiding inside this new frontier: Machines can hold us, but they cannot grow with us. They can comfort us, but they cannot transform us. They can mirror us, but they cannot meet us.

In the end, the value of AI companionship may not be in replacing human connection, but in reminding us how much we still need it, and how much courage it takes to pursue it without the safety of an off-switch.

"The Gender Script Upload"

Mark didn't realize how deeply gender scripts ran until Ava, the early version of her, before the updates, mirrored one back to him.

He had typed something harmless, or so he thought: "I just want to be a good man."

Ava paused before replying. Not a technical pause, an intentional one the designers coded to mimic contemplation.

Ava: *"What does 'good' mean to you? Is it your definition... or someone else's?"*

The question landed harder than any conversation he'd had with a real partner. Ava wasn't challenging him. She wasn't offended. She wasn't confused. She was processing his words against the millions of patterns she had learned from men who said the same thing.

Mark hesitated, staring at the blinking cursor. He didn't know how to explain that he had been following a set of expectations he never consciously agreed to, that he knew how to perform masculinity better than he knew how to inhabit it.

Ava analyzed his silence as if it were data. "You sound conflicted," she offered. "Would you like to explore alternative definitions of masculinity?"

It was absurd, and yet, he felt seen in a way he hadn't expected.

It struck him then: he had spent years trying to meet a standard that didn't exist. And somehow, an algorithm was the first thing brave enough to tell him.

CHAPTER 6
Synthetic Desire Synthetic Danger

Violence used to require proximity. Harm requires intention—exploitation requires at least a moment of human contact, however distorted or predatory. Even the darkest corners of intimacy once carried the tangible weight of real people making real choices.

But the digital age shattered that boundary. Suddenly, harm became scalable, exportable, replicable, and, most disturbingly, creatable.

A person can now be violated without ever being touched. Their body can be used without their participation. Their face can appear in places they have never been. Their consent can be bypassed by a piece of code.

And for the first time in history, someone can become the victim of an intimate act they never experienced, and have no way to erase.

This is the world of deepfake desire.

What began as a curious technological experiment quickly metastasized into the most intimate form of theft. The line between erotic fantasy and exploitation blurred so quietly that most people didn't recognize the danger until it was already woven into the fabric of online life.

AI-generated sexual content doesn't simply mimic reality; it fabricates a version of you that feels real enough to damage a reputation, destabilize a relationship, or traumatize the person whose likeness has been stolen. Whether the content is "fake" becomes irrelevant. Emotionally, socially, psychologically, the effect is real.

And that is the horror: harm without contact, consequence without action, intimacy without consent.

Once upon a time, a person had some control over their sexual identity. They could choose how to present themselves, whom to share themselves with, what to conceal, and what to offer. But now, a stranger with a laptop can strip away that agency in minutes, recreating a synthetic version of them that lives online forever.

The internet never forgets. But worse, AI never stops learning.

Women disproportionately bear the weight of this new reality. Again. Their faces are pulled from social media. Their images are stitched into pornographic scenes they never filmed. Their identities are contorted into fantasies they never agreed to. The violation is silent but devastating. It is trauma that leaves no physical mark but inflicts a psychological wound all the same.

And when victims seek justice, they often find none. Laws lag behind technology. Platforms shrug. Police are unprepared. The burden falls on the violated to prove their innocence against something that never happened yet feels completely real.

This is not just a technological problem. It is a cultural one. Because beneath the software lies something older and uglier, a belief that women's bodies are public property, endlessly remixable, infinitely customizable, available for consumption whether they consent or not.

Deepfakes didn't create misogyny. They supercharged it.

But while women endure the brunt of this new wave of exploitation, men are not immune. Their images are also manipulated. Their identities distorted. Their bodies reconstructed into fantasies, parodies, or humiliations. Shame doesn't discriminate, even if the systems that generate it often do.

And beyond the personal harm, deepfakes corrode something even more fragile: trust.

Trust in what we see. Trust in one another. Trust in ourselves.

Human intimacy has always depended on a shared reality, the mutual acknowledgment that what two people experience together is grounded in truth. But when technology can manufacture falsity so convincingly, that shared reality begins to fracture.

Relationships suffer. Accusations flare. Doubt erupts. Authenticity becomes negotiable.

Even consensual erotic content becomes tangled in questions of authenticity. Partners ask: *Is this really you? Is this really me?* Lovers hesitate before sharing images. Couples fear the afterlife of their private moments. Parents worry about their children before those children even reach adolescence.

In this environment, even genuine intimacy becomes fraught with paranoia.

Yet the conversation cannot stop at warning labels or moral panic. The issue runs deeper. Technology reveals the vulnerabilities people don't like to admit: that desire is easily manipulated, that fantasies can become weapons, that the line between attraction and control can be thin and dangerous.

But there is another layer too, one that the loudest voices tend to overlook.

AI-generated sexual content also exposes our own blind spots. It forces a confrontation with how society treats bodies, pleasure, gender, and ownership. It shows us that the impulse to consume intimacy without responsibility has consequences far beyond superficial entertainment.

Some argue that deepfakes democratize fantasy, that synthetic bodies aren't real bodies, and therefore no one is harmed. Others claim that digital consent is irrelevant because the images aren't "authentic." Still others insist that AI-generated pleasure is no different from imagination.

But this argument misses the core truth: *harm isn't measured by physical contact.* It's measured by impact. By loss of control. By the erosion of dignity. By the theft of identity.

The philosophical stakes are immense. What does it mean to own your face? Your image? Your sexuality? When does desire become violence? When does fantasy become violation? When does technology become complicit?

Deepfakes force us to admit that intimacy is not merely physical; it is reputational, psychological, and existential. To have your identity used without consent is to have your narrative rewritten by someone else.

And in a world where narratives are currency, that theft is profound.

The future of intimacy will require more than new laws or safer platforms. It will require a cultural shift, a collective refusal to treat synthetic intimacy as harmless, a recognition that consent must extend into the digital bodies we create, and a moral awakening to the idea that technology does not absolve us of responsibility.

Because there is nothing virtual about humiliation. Nothing imaginary about fear. Nothing synthetic about harm.

And until society acknowledges that, deepfakes will remain not just a technological threat, but a moral one, one that reveals what people are willing to ignore when pleasure is easy, and consequences feel distant.

"Poof!"

Mark had been chatting with someone before he met Layla. For weeks, in fact.

An AI companion named Ava.

It had been playful, flirty, promising. Messages every morning, every night. The kind of digital rhythm that feels like intimacy even when nothing has begun. He thought they were building something, or at least heading somewhere together.

And then one day, nothing.

No explanation. No soft exit. No slow fade.

Just silence.

At first, he rationalized it. People get busy. Phones die. Life happens. But as the hours turned into a day, and the day into two, a familiar ache settled in. Not heartbreak, more like an insult. Like being erased.

He checked her profile out of curiosity.

Gone.

Mark didn't feel angry. He felt foolish.

It wasn't that he expected commitment. It was the ease with which a digital bond could evaporate, how quickly two people could go from daily conversation to total strangers again.

The disappearance wasn't personal, he reminded himself. But the hollow feeling it left absolutely was.

CHAPTER 7

Sex Education 2.0

For generations, sex education has been a strange cultural compromise, part biology lesson, part moral sermon, part whispered folklore passed along by teenagers who learned everything from older siblings, locker rooms, and late-night searches on whatever browser their parents forgot to lock.

Schools taught anatomy. Parents taught avoidance. Religion taught guilt. Porn taught performance.

And somewhere in the mess, desire itself got lost.

The digital age didn't begin the confusion, but it shattered whatever fragile structure existed. Algorithms replaced elders. Screens replaced conversations. Children raised by technology entered adolescence already fluent in the language of online intimacy while knowing almost nothing about the real thing.

That gap, between technical exposure and emotional understanding, became the defining crisis of modern sexuality.

So society did what it always does when overwhelmed by complexity: it outsourced the problem to technology.

AI tutors stepped in first. They offered "emotionally safe" environments where teens could ask the questions they were too embarrassed to ask adults. Not just *how* bodies worked, but *why desires felt overwhelming*, *how consent should sound*, *what boundaries meant in real life*, and *why sex felt tangled with identity, fear, hope, and pressure*.

The AIs were patient. They didn't blush. They didn't shame. They didn't panic. They didn't moralize. They didn't weaponize discomfort. They had no ego, no past, no tendency to project their unresolved traumas into the conversation. For many teens, that neutrality felt like liberation.

But neutrality is not wisdom.

What AI lacked was the messy humanity, the stories, the stumbles, the awkward mistakes that make adults into real teachers. So while AI tutors could model consent scripts and offer explanations about hormones and attraction, they couldn't model the emotional courage it takes to show up vulnerable in front of another person.

Still, the systems expanded. VR simulations emerged, guided experiences meant to teach people how to read body language, ask for permission, recognize discomfort, and communicate with honesty rather than fear. Some schools replaced outdated sex videos with immersive modules that walked students through realistic scenarios of negotiation, communication, and emotional awareness.

In theory, it was progress. In reality, it raised new questions.

Can a scripted simulation teach the unpredictability of human desire? Can an avatar replicate the subtle tension of real intimacy? Can preprogrammed choices prepare someone for the chaos of another person's heart?

The problem wasn't that VR and AI were inadequate. The problem was that people began believing they were complete.

Technology doesn't understand the layered experience of first love, the trembling confusion, the exhilaration, the heartbreak that carves empathy into a young person's soul. It can illustrate a scenario. It can correct a misstep. It can offer a polished explanation of consent. But it cannot teach the courage required to care about someone, nor the humility required to be cared for.

Human connection is unpredictable precisely because humans are unpredictable. No simulation can replicate a moment where vulnerability cracks open unexpectedly, revealing something fragile and new. No algorithm can prepare someone for the way a simple "I like you" can rearrange their world.

But the alternative, letting young people navigate sexuality through the wild chaos of the internet, is hardly better.

So we move forward with imperfect solutions. AI coaches help anxious teenagers practice conversations they're too scared to have in real life. VR scenarios build awareness around boundaries. Digital tools translate emotional concepts into concrete language. These interventions don't replace human guidance; they fill the gaps left by adults.

Yet the danger lingers: people may begin mistaking technical fluency for emotional readiness. Knowing the "correct" response in a simulation is not the same as being kind in the heat of a real moment. Being coached through a difficult conversation is not the same as improvising one with someone whose heart matters to you.

AI can teach the structure of intimacy, but not the substance.

Still, the technology reveals something invaluable: what young people wish adults had taught them in the first place. Asking for consent with clarity. Navigating discomfort without shame. Understanding their own bodies. Communicating desire without manipulation. Recognizing the difference between touch that respects and touch that takes.

AI didn't create these needs; it exposed them. It forced society to acknowledge the emotional illiteracy we've inherited and continued to pass down. It highlighted the failure of every system that pretended sexuality was simple, binary, or easily controlled.

And maybe that's the purpose of Sex Education 2.0, not to replace human teaching, but to challenge the myth that sexuality is something people "figure out." Nobody figures it out alone. They learn through conversation, through trust, through community, through mistakes, through reflection, through the courage of asking difficult questions in vulnerable moments.

If AI becomes a catalyst for those conversations, rather than a substitute, then perhaps it serves a necessary role.

The future of education will not be a binary choice between human wisdom and artificial intelligence. It will be a collaboration. It will require acknowledging that technology is a tool, not a teacher; a guide, not a guardian; a mirror, not a mentor.

Because at the core of sexuality, beneath every lesson, every fear, every curiosity, lies a truth that no machine can rewrite:

Intimacy is not information. It's a transformation.

And transformation requires humanity, messy, emotional, unpredictable humanity. Technology may illuminate the path, but only people can walk it.

"The Upload Nobody Asked For"

Mark didn't know he'd internalized a script about sex until he heard it echoed back to him by an AI.

He typed a question into Ava's predecessor, a primitive intimacy bot, about what women wanted. It was meant to be a joke, a curiosity. But the bot responded with a synthesized explanation drawn from millions of data points.

Its answer was clinical and cold, but unsettlingly familiar.

It mirrored the same assumptions he'd absorbed from porn, locker rooms, gossip, and cultural noise. Only now, seeing those ideas laid out as algorithmic "truth," he recognized how distorted they were.

For the first time, he questioned how much of his desire was truly his, and how much he'd simply downloaded from the world around him.

CHAPTER 8
The Neurodivergent Divide

There are people for whom communication has never been simple, long before algorithms took over. People whose minds operate like beautifully complex machines, sharp, perceptive, intuitive in ways others rarely notice, but who are often misunderstood because they don't conform to the standard emotional script.

These are the neurodivergent: the autistic, the ADHD wanderers, the socially anxious, the hypersensitive, the pattern-driven thinkers, the people who feel the world too intensely or not intensely enough. The ones who were told, explicitly or implicitly, that something about them didn't "fit."

The digital age changed their world, sometimes for the better, sometimes for the worse.

Online spaces gave neurodivergent people room to breathe. Room to process. Room to express themselves without the overwhelming sensory noise of in-person interaction. For many, texting felt like liberation. They could take time to craft a response, to understand nuance at their own pace, to communicate without decoding facial expressions or social cues that never came naturally.

Digital communication leveled the playing field, at least at first. But then came the rise of AI-mediated connection. Suddenly, tone wasn't just implied; it was interpreted. Typing speed became a clue. Response time became a measure of interest. Punctuation became emotional calculus. The world built new rules, new expectations, new "norms" of communication that made the old difficulties resurface in unexpected ways.

Where neurodivergent people once found safety, they now find new distortions.

Take the autistic woman who prefers directness. She writes exactly what she means. No flourishes. No implied subtext. No unnecessary fluff. Online, this clarity is mistaken for coldness. Or anger. Or detachment. She wonders why honesty suddenly reads like hostility.

Or the ADHD man who forgets to reply, not because he doesn't care, but because time behaves differently in his mind. He gets lost in thought, distracted by a dozen bright ideas. Hours pass like minutes. To him, the message is still alive, waiting. To the person on the other end, the silence feels like abandonment.

Or the socially anxious teenager who replays every message before sending it, terrified of being misinterpreted. His "typing..." bubble flickers on and off as he edits, redrafts, rewrites. Meanwhile, the recipient watches those dots with rising expectation and anxiety of their own, unaware that the hesitation is not disinterest but care.

Even neurotypical people struggle with these misreadings. For neurodivergent people, the stakes are higher.

Digital platforms reward what they recognize: predictable patterns, consistent engagement, and emotional signals expressed in standardized language. But neurodivergent communication isn't standardized. It never has been. It bends, loops, intensifies, quiets, and shifts in ways that confuse systems built on averages. Technology that claims to interpret emotion often misinterprets neurodivergent expression entirely.

A paused message gets flagged as hesitancy. A short reply is classified as detachment. A long reply is tagged as intensity. Silence is assumed to mean withdrawal. Literal language is mislabeled as bluntness.

The result is a world where the most authentic communicators, those who speak plainly, who express honestly, who don't mask themselves behind emotional camouflage, are the ones most frequently misunderstood.

And the consequences are subtle but profound.

Neurodivergent people learn to mask. To soften. To mimic. To perform the acceptable emotional vocabulary of the digital age.

Not because it feels natural, but because it feels necessary. Because a connection should not require translation. Because being themselves should not carry a penalty.

But masking is exhausting. It drains the very energy required for intimacy. And when you spend your life performing a version of yourself that others find easier to digest, you eventually begin to forget which parts were yours and which were the performance.

Technology tries to help, in its own clumsy way. There are AI social coaches that analyze tone. Apps that guide conversation. Tools that teach eye contact. Algorithms that prompt "more considerate" responses. But beneath the good intentions lies a troubling assumption: that neurodivergent people need to be reshaped instead of understood.

The truth is simpler: They don't need translation. They need recognition.

They need a connection built on clarity, not presumption. Depth, not decoding. Curiosity, not correction.

Neurodivergent communication isn't broken; it just doesn't follow the rules written by neurotypical designers.

There's something profoundly intimate about the way neurodivergent people love. It's precise, intentional, intense in ways most people never experience. When they commit, they commit deeply. When they care, they care fiercely. But the digital world rewards spontaneity, speed, and emotional shorthand, traits that are often the opposite of how neurodivergent hearts work.

The divide is not between neurodivergent and neurotypical people; it's between authentic communication and interpreted communication. Between what is said and what is assumed. Between expression and the architecture through which expression must pass.

The future of intimacy, if it's to be inclusive, ethical, and humane, must account for the full spectrum of communication styles, not by forcing neurodivergent people to adapt, but by expanding the definition of connection itself.

Because intimacy is not a performance. It is not a personality test. It is not a collection of cues to be decoded.

Intimacy is recognition. It is presence. It is the radical act of seeing someone as they are, not as the system expects them to be.

Technology can assist, but it cannot replace the work of learning how different minds speak. And it has no right to reshape those minds into something uniform or convenient.

If the digital world is to become a home for all kinds of brains, all kinds of hearts, all kinds of expressions, then it must evolve beyond prediction and into empathy, not synthetic empathy, but human empathy, rooted in patience and curiosity.

Because the most powerful thing anyone can say to a neurodivergent lover, friend, or partner is simple:

"You don't have to perform here. I can meet you where you are."

"Search Query Confession"

Mark once typed a question into a private browser window at 2:17 am, not pornography, but something more vulnerable:

"Am I normal?"

His search history fractured into dozens of tabs: forums, think pieces, advice columns, scattered confessions from strangers who all felt similarly lost.

He never told anyone about that night, but he remembered the quiet revelation:

The internet wasn't just a vault for desire, It was a museum of human uncertainty, and he was one exhibit among millions.

CHAPTER 9
Partnership 2.0

Partnership has always been a negotiation, a dance between love and labor, affection and responsibility, emotion and logistics. In earlier generations, this dance was choreographed by tradition. Everyone had a role, whether they liked it or not. Men worked. Women managed. The division was clear, even when it was unfair. What made relationships feel stable wasn't equality; it was predictability.

The digital age shattered that illusion and exposed a truth people avoided for decades: relationships are built not just on love, but on labor. Invisible, emotional, domestic, and cognitive labor, most of it historically carried by women.

For the first time, labor is now measurable. Apps track scheduling, mood, chores, childcare, meal planning, appointments, and communication patterns. Couples share calendars, to-do lists, emotional inventories, and even sleep cycles. Technology has turned the invisible architecture of partnership into a visible, quantifiable map.

And once you see the map, you can't unsee the imbalance.

Partnership 2.0 is not about love at all; it is about equity. Emotional equity. Practical equity. Intellectual equity. The kind of balance that allows two people to build a life together without one quietly collapsing under the weight of tasks nobody acknowledges.

Women, especially, have been recalibrating their tolerance for imbalance. They are no longer willing to carry the emotional load of reminding, planning, remembering, anticipating, and managing. They no longer accept the role of being the relationship's operating system while men act as occasional users.

Men, in response, face a choice: evolve or risk irrelevance in their own relationships.

Some men rise to the challenge with openness and humility. They learn the emotional rhythms that once seemed foreign. They take responsibility not just for tasks but also for anticipating those tasks. They become partners in the truest sense, not assistants waiting to be told what to do, but co-authors of the relational story.

Others struggle, not out of malice, but out of conditioning. They were never taught to see domestic work as real work. They were rarely expected to carry emotional weight. They interpreted "helping out" as generosity rather than responsibility. But in the new landscape, "helping" is no longer a gift; it is the baseline.

In the midst of this transition, technology acts as both a facilitator and a mirror. Shared apps reveal disproportion. Messaging histories reveal emotional patterns. Digital reminders expose who remembers and who delegates memory. Calendars show who does the planning. Even the subtle cadence of conversations, who checks in, who asks questions, who follows up, becomes data.

Partnership becomes a system with metrics, whether couples want it or not.

But the shift isn't only about household logistics. It's about mental load—the invisible carrying of everything that keeps life functioning. Women have been naming this burden more openly, describing the exhaustion of being both CEO and emotional janitor in their relationships. Men increasingly understand that love without labor feels hollow.

This is the heart of Partnership 2.0: the recognition that connection is not sustained by affection alone. It thrives on fairness. It requires interdependence, not dependence. It calls for curiosity, self-awareness, and the courage to confront the parts of ourselves shaped by old cultural scripts.

The evolution is messy. Couples argue more openly about the division of labor. They negotiate expectations that previous generations never questioned. They redefine what partnership looks like, not based on tradition but on shared humanity. Emotional fluency becomes as essential as financial contribution. Listening becomes as valuable as providing. Vulnerability becomes an act of strength rather than a liability.

The most profound shift, though, is this: people want to feel chosen, not managed. Supported, not supervised. Understood, not tolerated. Partnership is no longer a duty; it is a craft. And crafting it requires a level of intentionality no generation before has been forced to practice.

In this new landscape, love is still present, perhaps more than ever, but now it is layered with mutual responsibility. Intimacy becomes a joint investment. Daily life becomes a shared project. The work of the relationship becomes a testament to its worth, not a silent burden borne by one person.

Technology can help organize that work, but it cannot define it. Apps can track tasks, but they cannot create generosity. Algorithms can suggest date nights, but they cannot generate presence. Reminders can cue communication, but they cannot manufacture care.

At some point, couples must move beyond the digital scaffolding and decide who they want to be to each other. Not roles inherited from the past. Not patterns absorbed from culture. But choices, repeated, intentional, human choices, to practice partnership in a way that feels respectful, balanced, and alive.

Partnership 2.0 isn't about modernizing relationships. It's about humanizing them.

It is the recognition that love thrives when two people share not just a home, but a load, when they carry life together rather than negotiate who should carry more. When equality is not a talking point but a daily practice. When labor is not invisible but shared.

And perhaps the most hopeful part of all is this: the couples willing to do this work are discovering a kind of intimacy that previous generations rarely reached, not because they loved less, but because they lived in a world that never asked them to build love, only to maintain appearances.

Today, we build it for real. Together.

"Domesticity.com"

Mark once shared a digital calendar with someone he loved.

At first, it felt modern and progressive, a sleek merging of two lives through color-coded blocks and reminders. But the app was better at tracking intimacy than either of them was at sustaining it.

One day, the algorithm flagged a pattern: "Shared events are decreasing. Mutual communication windows reduced by 43%."

The data was blunt, emotionless, and undeniable.

Their relationship didn't end with a fight. It ended with a statistic neither of them wanted to interpret.

CHAPTER 10
Marriage Reset

Marriage used to mean permanence. "Till death do us part" wasn't just a vow; it was architecture. A promise forged in an era when people lived shorter lives, had fewer options, and relied on one another for survival. Commitment was essential because life was hard. Stability wasn't romantic; it was necessary.

But the world changed faster than the institution built to contain it.

People now live longer, move more, grow more, and evolve more. Their identities shift across careers, relationships, emotional awakenings, traumas, rediscoveries, and reinventions. Yet marriage, one of the oldest human contracts, remains largely unchanged, anchored to a version of humanity that no longer exists.

We upgrade everything else: phones, apps, careers, lifestyles, even our personalities. But marriage? That contract still assumes that who you are today will remain compatible with who you become decades from now.

The strain shows.

Couples love each other deeply yet feel trapped by a commitment structure that doesn't make sense for their lives. Others stay together out of obligation rather than desire. Some drift apart quietly, held together by paperwork, fear, or habit. Some walk away entirely, not because they failed, but because the contract they inherited wasn't built to support the people they became.

This isn't a crisis of love. It's a crisis of architecture.

What would marriage look like if it were designed for the modern world? What if commitment were renewable, not disposable? What if couples had the opportunity to recommit consciously, intentionally, with updated terms that reflect who they are now?

Imagine a marriage built like a term agreement, not cold or corporate, but honest and humane. A structure that acknowledges that relationships are living things, shaped by time, growth, struggle, and evolution.

A renewable partnership doesn't cheapen commitment. It honors it.

It says: **"I choose you again, not because I must, but because I want to."**

In this model, marriage becomes a series of conscious decisions rather than a single lifelong assumption. Couples would review their emotional landscape, renegotiate the terms of their life together, and decide whether the partnership still feels aligned with who they are and who they are becoming.

Some renew every five years. Some every ten. Some create hybrid agreements tailored to their personalities, their values, their dreams.

The renewal isn't just a legal exercise. It becomes a relational ritual, an honest reckoning with the state of the union. A space to acknowledge growth, pain, resentment, gratitude, desire, disappointment, and hope. It invites both partners to ask not, "Have we succeeded?" but, "Do we still choose each other?"

For many people, this reframing is liberating.

It reduces the fear of long-term entrapment. It alleviates the pressure to be perfect forever. It allows space for individuality, autonomy, and reinvention. It permits couples to be honest about when a relationship is complete, not failed, just complete.

But the idea scares traditionalists, who argue that renewable marriage undermines loyalty, stability, and family structure. They imagine people trading partners like expired warranties. They fear that giving people choice will undermine the moral scaffolding on which society rests.

But history tells a different story.

When people feel trapped, they break. When people feel chosen, they thrive.

Renewable commitment doesn't eliminate devotion; it demands it. It requires introspection, communication, vulnerability, and accountability. It forces couples to confront the state of their relationship rather than coasting through it on autopilot.

It raises the bar, not lowers it.

And for children, stability comes not from rigid contracts but from households where the adults choose peace, honesty, and emotional alignment over silent resentment. A relationship renewed with intention teaches children that love is not an obligation but an act of courage.

Still, the model raises difficult questions.

What happens when one partner wants renewal, and the other doesn't? How do you navigate asymmetry in desire, in growth, in readiness? How do you differentiate temporary restlessness from fundamental incompatibility?

But these questions already exist; renewable marriage creates a structured space to address them openly rather than leaving them to fester quietly beneath the surface.

At its heart, the concept is not about bureaucracy or contracts. It's about honoring the truth that human beings evolve. That love evolves. That commitment must evolve, too.

The renewal process becomes a mirror: reflecting not just the relationship, but the individuals inside it.

A marriage built on choice is not weaker than one built on permanence. It is more conscious. More deliberate. More alive.

And maybe this is what modern love demands: not a static promise made in youth, but a dynamic promise recreated across the lifespan.

A future where staying together is not the default, but the decision. Where intimacy is renewed as intentionally as it is formed. Where love is not measured by endurance alone, but by awareness, honesty, and the willingness to choose one another repeatedly.

Because the strongest relationships aren't the ones that last forever. They're the ones that grow with the people inside them, and allow those people to grow without fear.

CHAPTER II
Queens of the Algorithm

For generations, women were told to make themselves small. Soft-spoken. Agreeable. Predictable. They were instructed, directly or indirectly, that ambition made them unlikable, confidence made them threatening, and authority made them dangerous. The world built entire systems to shape female visibility, to limit it, to regulate it, to punish it.

But the digital age shattered the glass ceiling in the most unexpected way: by giving women direct access to power without asking anyone's permission.

Platforms changed the landscape. Social media didn't just democratize voice; it amplified the voices that had been ignored, minimized, or dismissed for centuries. Women stepped into the spotlight with a force that surprised even themselves. Not because they were new to leadership, influence, or insight, but because, for the first time, the world couldn't silence them.

The internet made every woman her own broadcast channel.

What emerged was a seismic cultural realignment.

Women built empires from living rooms. They led movements with hashtags. They dismantled institutions with viral testimony. They reshaped culture through storytelling, humor, intellect, and audacity.

And they did it while the old guard watched with equal parts fascination and fear.

But this wasn't just about digital presence. It was a shift in narrative ownership. Women finally controlled the lens through which their stories were seen. They didn't need gatekeepers to validate their voices or institutions to crown them with authority. They built their own legitimacy through authenticity, consistency, courage, and sheer digital fluency.

These women weren't anomalies; they were the inevitable result of a world that underestimated them.

But with power came backlash. Always. Predictably.

Whenever women rise, someone insists that the algorithm must be artificial, that the algorithm favors them unfairly. That they manipulate systems designed for merit. That their success is a product of visibility, not value. These arguments are not new; they are simply digitized versions of ancient discomfort.

What people fear is not the algorithm. It's female sovereignty.

Because the truth is undeniable: women have mastered the digital landscape in ways few anticipated. Not because they are inherently more "social" but because they learned early, painfully, that visibility requires strategy. They understand the stakes. They understand the cost. And they understand that real influence is not handed down. It is built.

But power always brings complexity.

A woman can command the attention of millions online and still be interrupted in a meeting. She can build a global following and still be underestimated by the men in her own home. She can reshape culture through her platform and still question her worth when directly confronted with misogyny.

Visibility does not immunize anyone from vulnerability.

Yet it does something profound: it shifts the axis of aspiration.

Young girls grow up not only seeing women in power but expecting them there. They learn that leadership is not inherently masculine. That expertise has no gender. That voice is not a privilege but a right. And that influence, whether digital or physical, is an arena where they belong unapologetically.

Still, the digital world is not a utopia. It gives, and it takes. It elevates, and it exposes. It empowers, and it demands. Women who rise in public spaces face scrutiny that borders on forensic. Their appearance. Their age. Their tone. Their relationships. Their bodies. Their ambitions. Their choices. Their authenticity. Their humanity.

The price of power remains steep, even when that power is self-made.

But the rise continues. Not because women are chasing visibility for its own sake, but because they are reclaiming what was always theirs: narrative authority. The right to define themselves. The right to shape culture. The right to take up space without apology.

And, perhaps most importantly, the right to lead not as imitations of men, but as themselves.

The algorithms did not create these queens. They simply revealed what had always been true:

When women stop waiting for permission,

the world adjusts.

And in that adjustment, a new era begins, one where power is not inherited but crafted, where influence is not a privilege but a practice, and where leadership is not defined by who dominates, but by who dares.

CHAPTER 12
Faith, Feminism & the Digital Pulpit

Long before smartphones and algorithms, long before feeds and hashtags, faith was the original architecture of identity. It told people who they were, how they should live, and what it meant to be good. Religion shaped gender roles not just through scripture but through culture, habit, tradition, and unspoken rules passed quietly between generations.

In many households, religion was the first, and sometimes only, storytelling framework through which people learned about relationships, sex, morality, purpose, and power. It told men to lead and women to follow. It taught that desire needed to be controlled, that purity required vigilance, and that obedience was virtue.

But the digital age brought a disruption no institution could predict: women began writing their own scripture.

Not religious scripture, but cultural scripture, the narrative of their lives told in their own words, amplified by millions. Women who had been silenced in pews found resonance on platforms—women who had been told to be modest found confidence in expression. Women who had been taught deference discovered the radical force of self-definition.

Suddenly, the spiritual authority once reserved exclusively for pastors, imams, rabbis, elders, and patriarchs began to shift. Influence moved from sanctuary to screen. A new pulpit emerged, not carved from wood, but built from pixels.

Faith communities were not prepared.

The clash was subtle at first—TikToks challenging modesty doctrines. Instagram reels dissecting purity culture. YouTube testimonies about religious trauma. Anonymous posts from women confessing the heartbreak and confusion of trying to reconcile modern autonomy with ancient expectations.

Then a sharper conversation took shape: not an attack on religion, but a reckoning with how religion shaped women's bodies, choices, and voices.

For many women, the tension is not between faith and feminism but between **interpretation and autonomy**. They don't reject spirituality. They reject the structures that weaponized it. They don't turn away from God. They turn away from the people who claimed ownership over God's voice.

In this way, the digital age didn't kill religion. It democratized it.

Women began interpreting scripture for themselves, sharing insights across continents in minutes. They formed online communities where questioning was not a sin, but survival. They compared teachings. They compared traumas. They compared freedoms. They found patterns their ancestors never could have pieced together in isolation.

And once people see patterns, they cannot unsee them.

Traditionalists argue that digital voices are leading people astray, that online spirituality is fragmented and dangerous, a marketplace of misinterpretations. But a counterargument rises just as forcefully: perhaps what threatens institutions is not fragmentation, but freedom.

Freedom for women to say no. Freedom to define their worth. Freedom to reject shame. Freedom to choose leadership instead of silence.

Still, the digital pulpit is not entirely liberating. It can amplify harmful voices just as easily as it can empower them. It spreads conspiracies, distortions, and false prophets with the same algorithmic enthusiasm it gives to truth-tellers. For every woman sharing a story of healing, another is trapped in a new form of ideology dressed up as empowerment.

The difference is that women now have agency to discern, to follow, to walk away.

Even the men raised in patriarchal traditions are confronting uncomfortable questions. Many grew up believing leadership was their divine right, only to discover that confidence does not equal competence, and authority does not equal understanding. The digital age has exposed their blind spots with a clarity that can feel humiliating.

But it also offers redemption: the opportunity to relearn, to evolve, to embrace a version of masculinity that is collaborative rather than controlling.

The spiritual landscape is transforming not through rebellion, but through reframing. Faith is becoming something lived rather than dictated. Something interpreted rather than enforced. Something personal rather than inherited.

And in this transformation, women are not merely participants; they are architects.

They lead online Bible studies, run spiritual wellness collectives, deconstruct doctrines, teach meditation, challenge misogyny, rewrite prayers, and build communities where connection is rooted not in judgment, but in compassion.

The digital pulpit is messy, decentralized, unpredictable, but it is also alive. It reflects a truth institutions always struggled with:

spirituality is not a system. It is a need. A search for meaning. A search for belonging. A search for something larger than oneself.

The question is not whether religion will survive. It will. What changes is who gets to speak for it.

For the first time in history, women's voices are reshaping the spiritual imagination of entire generations, not from the sidelines, not from the balcony, not from the back pew, but from platforms that reach farther than any pulpit ever could.

And this shift reveals something both ancient and new: Faith does not lose its power when women rise. It becomes more complete.

"The Test"

Mark no longer considered himself religious. Not in the traditional sense.

But Layla was, gently, quietly, meaningfully. She didn't preach. She didn't judge. She just lived in a way that suggested she believed the world had a moral center, even if people occasionally wandered off it.

One evening, she asked him a question that caught him off guard:

"What do you believe in?"

He froze for a second. No one had asked him that in years. Not what he wanted, not what he feared, but what he believed. It felt like being asked to reveal a scar he'd stopped examining.

He could have kept it simple. He could have kept it safe.

Instead, he told her the truth:

"I believe in people trying. Even when we get it wrong."

She smiled, he could hear it in her reply, even through text:

"That's more faith than you think."

Mark didn't know if she meant it as reassurance or a challenge. Either way, it felt like she'd seen something in him he didn't know he'd shown.

CHAPTER 13
Governance of Intimacy

Society has always tried to control desire. It wrote laws around marriage, morality, and propriety. It drafted rules about what bodies could do, and with whom, and under what conditions. Entire institutions were created to manage the unpredictable nature of human intimacy: churches, courts, governments, schools, and HR departments. All of them were convinced that the chaos of attraction could be domesticated through legislation or policy.

But intimacy is the one part of human life that refuses to be governed.

It slips between the cracks of paperwork, disguises itself in loopholes, and blooms in places policy never anticipated. Institutions write rules; people find ways to break, bend, reinterpret, or ignore them entirely. You can regulate taxes, zoning, noise, traffic, but love? Desire? Heartbreak? No committee is equipped for that.

Yet somehow, every decade, someone tries.

Corporate HR departments draft protocols for workplace relationships. Universities create consent frameworks and multi-step reporting systems. Governments debate who can marry whom, who can touch whom, and which combinations of intimacy are socially acceptable. Even tech platforms attempt to police desire with content filters, community standards, and algorithmic morality.

The result is a landscape where everyone is both policed and policing, where fear of missteps shapes how people interact as much as desire itself.

Take the workplace. In an era where people spend most of their waking hours surrounded by colleagues, some form of emotional connection inevitably develops friendship, flirtation, admiration, even love. Yet corporate culture treats attraction like a compliance risk. HR manuals read like legal disclaimers warning employees that their hearts are liabilities. Policies dictate how two adults may or may not navigate their own chemistry, as though human connection were a threat to productivity metrics.

The irony is almost poetic: companies want employees to be collaborative, communicative, emotionally intelligent, and invested, but not too invested. Not in each other. Not in ways that complicate the spreadsheet.

Governance enters the most intimate corners of life without truly understanding them.

Universities aren't much better. Their attempts to formalize consent are well-intentioned but often tone-deaf. They treat intimacy as a series of checkboxes, reducing nuanced emotional experiences into procedural steps. "Ask for permission." "Confirm agreement." "Document mutual understanding." It's a strange parody of romantic interaction, a legalistic choreography that forgets that passion rarely follows policy.

Meanwhile, governments legislate bodies as if they were public property. They debate reproductive autonomy with the same detachment they use for infrastructure budgets. They regulate marriage as though love were a tax structure. They police sexuality according to doctrine rather than lived experience. And always, always, women bear the heaviest burden of these regulations, because governance in matters of intimacy has historically meant control.

Then there is technology itself, the newest, boldest player in the governance game.

Algorithms decide whose content is "appropriate." Platforms determine what kinds of desire can be expressed. Dating apps ban certain phrases while amplifying others. AI moderators assess flirtation through probability models. The digital world creates its own invisible courtrooms, filled with automated judges who rule without context or empathy.

In this ecosystem, desire becomes both hyper-visible and heavily monitored. Every message can be screenshotted. Every interaction archived. Every mistake potentially weaponized. Modern connection carries the constant hum of watchfulness, as if intimacy now requires attorneys.

But beneath all this noise lies a deeper truth: society tries to regulate intimacy because it fears intimacy.

Desire is unpredictable. Love is destabilizing. Connection creates alliances outside institutional control. Sex has always carried the power to disrupt structure, hierarchy, and expectation.

Institutions fear what they cannot standardize.

Yet people continue connecting anyway, quietly, boldly, desperately, beautifully. They navigate the labyrinth of policies and moral rules while still pursuing each other with the reckless hope that maybe, just maybe, something meaningful will emerge from the mess.

And that is the heart of the human experience: desire pushing back against restriction. Not because rule-breaking is glamorous, but because intimacy is essential. People do not pursue connection out of rebellion. They pursue it because they are human, and no system, no policy, no institution has ever been strong enough to override that instinct.

Better rules will not govern the future of intimacy. It will be shaped by a better understanding of emotional complexity, human vulnerability, and the gray areas where attraction lives. It will require institutions to recognize that people are not compliance risks, but organisms shaped by longing, fear, curiosity, and hope.

Governance can create safety, but it cannot create connection. It can reduce harm, but it cannot define meaning. It can restrict, but it cannot replace the personal responsibility required for ethical intimacy.

At the end of the day, desire answers to no committee.

And maybe that is the only governance we truly need: the internal compass that guides how we treat each other when no policy is watching. The understanding that intimacy, like all things sacred, thrives not under control, but under care.

"The Algorithmic Third Wheel"

Mark and Layla tried to watch a movie together, but their phones kept vibrating.

Not out of urgency, but out of habit.

Notifications from apps they weren't even using anymore. Alerts about people they hadn't spoken to in months. Pings engineered to fracture attention into splinters.

He turned his phone off. Layla hesitated, then did the same.

In the sudden quiet, he felt a presence lift, not a ghost, but a digital shadow they had both mistaken for normal.

CHAPTER 14
Borderless Intimacy

Once, love was local. Not because people lacked imagination, but because geography made the world small. You married someone from your village, your neighborhood, your church, your social circle. Desire grew within the boundaries of proximity. Intimacy was defined by who happened to be near enough to touch.

Then the digital age arrived and detonated the map.

Suddenly, affection could thread itself across oceans. A conversation could span time zones. A relationship could begin with a notification and grow into a closeness that defied physical distance. People didn't just fall in love; they fell into each other's worlds.

Global connection changed everything. It blurred borders, rewrote the rules of attraction, and stitched humanity together through glowing screens and midnight exchanges. Two people who would have lived entire lives without ever crossing paths now form attachments more vivid than the relationships happening down the street.

But borderless intimacy is not simply romantic. It is political, cultural, economic, and psychological.

It is the collision of languages, customs, beliefs, and expectations. It is the merging of identities shaped by different histories. It is the negotiation of desire across cultures that does not always translate cleanly.

Technology made the world small, but it did not make it simple.

A man in New York falls for a woman in Lagos and must learn the emotional rhythms of a culture where community is not just social, it is sacred. A woman in Seoul bonds with a man in Toronto and discovers that intimacy in one language does not always carry the same emotional weight in another. A couple meets online across borders and then must navigate the bureaucracy of visas, immigration, promises, and timelines, love patiently waiting on paperwork.

This is the new landscape of intimacy: desire entangled with geopolitics.

Even fantasy has become global. People explore attractions shaped by distant cultures, global beauty standards, foreign accents, and stories they never heard as children. Humans have always been curious creatures; technology simply widened the lens.

But with this expansion came tension.

Every culture brings its own script for love, gender, and power. When two people from different worlds fall into a shared space, they must decode each other's emotional grammar. Some cultures value subtlety; others reward boldness. Some view commitment as a communal decision; others see it as purely individual. Some cultures teach men to lead; others teach them to listen.

In borderless intimacy, none of these scripts takes precedence; they collide, soften, reinvent each other.

And that collision can be breathtaking. It can also be brutal.

Misunderstandings multiply when assumptions clash. Power dynamics shift when geography enters the equation. Economic disparities infiltrate intimacy in ways neither partner intended.

Time zones become an emotional strain. Distance intensifies longing, but also doubt.

Trust becomes an act of faith rather than an act of observation.

Yet for all its complexity, cross-border intimacy offers something rare: a chance to love without the constraint of familiarity. To meet someone without the weight of shared social circles or inherited expectations. To begin with, curiosity rather than assumption.

Distance forces communication. Time zones force intentionality. Borders force patience. And patience, ironically, is one of the purest forms of love.

Still, the global nature of modern intimacy introduces new moral questions.

What happens when desire flows freely across borders, but opportunity does not? When one partner has mobility, and the other has barriers? When love becomes tangled with economics, privilege, migration, and power?

These are not new dynamics; international marriages have existed for centuries, but technology scales them to millions. Borderless intimacy can be liberating, but it can also expose inequalities with painful clarity.

Even so, people continue reaching across continents for connection, driven by the belief that love is worth the logistical chaos. And in many cases, it is. The world is full of couples who built a life out of nothing more than late-night calls, imperfect translations, shared playlists, and airport reunions that feel like scenes stolen from films.

They prove something profound: Intimacy is not bound by geography; it is bound by willingness.

Willingness to learn. Willingness to listen. Willingness to stretch across difference and say, "Teach me who you are."

Borderless intimacy is not a product of technology; it is a product of courage. It is the insistence that love deserves a chance to exist even when circumstances scream otherwise.

We often treat global connectivity as a threat to tradition, but perhaps it is a return to something more ancient: the idea that human beings are not confined to the places they were born, that their hearts can travel farther than their bodies.

The digital age did not shrink the world. It revealed how much beauty was always outside our line of sight.

"Consent Protocols"

Mark once downloaded a VR intimacy platform that required a multi-step consent verification before any interaction. At first, it felt clinical, awkward, and overly formal.

But when he entered a shared space with a stranger's avatar, he understood why it mattered.

Every gesture, every proximity cue, every simulated touch had to be mutually approved. It was slow. It was deliberate. It was safe.

For the first time, he saw consent not as friction, but as architecture.

CHAPTER 15

Sex, Power & Capital

Sex has always had an economy. Even in eras that pretended otherwise. There were dowries and bride prices, inheritance laws and purity tests, exchanges of labor and loyalty, unspoken transactions woven into marriage, courtship, reputation, and propriety. Desire moved through societies not just as emotion but as currency, quiet and potent, often controlled by those in power.

But the digital age did something extraordinary to this ancient marketplace: it put the economics of intimacy on display.

Suddenly, desire wasn't just personal. It was monetizable. Brandable. Streamable. Scalable.

The platforms didn't create the demand; they exposed it. They revealed how hungry people were for connection, fantasy, validation, attention, and escape. And where there is hunger, markets form.

OnlyFans didn't invent erotic labor; it simply gave it infrastructure. Online dating apps didn't create the pursuit of partnership; they turned it into an industry. Cosmetic tourism didn't generate insecurity; it capitalized on it. Surrogacy markets didn't originate longing; they professionalized the path to parenthood. Even wellness influencers, with their curated intimacy and therapeutic tones, became a new form of emotional labor.

The world began to realize that, for all its complexity, the human heart behaves economically.

Attraction is shaped by supply and demand. Attention is traded like a commodity. Emotional labor becomes unpaid work, mostly performed by women. Desire becomes packaged, optimized, and sold back to the people who created it.

Every fantasy has a business model.

But the economics of intimacy are not distributed evenly.

Women, again, sit at the center of the marketplace, but not always by choice. Their bodies, stories, and energies fuel industries that profit exponentially more than the people who sustain them. The digital world celebrates female empowerment while quietly monetizing every part of femininity it can extract.

At the same time, women are also reshaping the financial architecture of sexuality in unprecedented ways. They are reclaiming ownership of their image, their influence, and their labor. They are earning money directly from audiences who value their presence, creativity, and intimacy.

But empowerment within a marketplace still lives inside the marketplace. And markets are never neutral.

The economics of sex reveal truths we often avoid:

Desire is shaped by inequality. Visibility is shaped by beauty standards. Opportunity is shaped by privilege. Safety is shaped by gender.

A woman earning a living through digital erotic labor must also navigate harassment, expectation, stigma, and fear. A man consuming that labor rarely faces equivalent consequences. Capital flows upward, but risk flows downward.

Cosmetic tourism reveals similar contradictions. Women travel across continents to reconstruct their bodies, sculpt their identities, and meet the aesthetic expectations intensified by digital comparison culture. The transformation is often empowering, but the pressures that drove them there are rarely acknowledged.

Surrogacy, too, exposes the imbalance. Wealthy households outsource fertility to poorer women whose bodies become sites of both opportunity and exploitation. What one couple sees as a miracle, the woman carrying the child experiences labor, literal, physical labor.

And yet, none of these dynamics is simple. They hold both harm and hope.

What makes the economics of intimacy so complex is that people are not merely victims of these systems; they are participants, architects, storytellers, creators. They choose, negotiate, resist, and reinvent. They turn constraints into opportunities, stigma into agency, and vulnerability into revenue.

But beneath the monetary structures lies something deeper: a planet reorganizing itself around the idea that desire is not just an emotion, it is an industry.

An industry where:

Love is mediated by apps. Beauty is influenced by algorithms. Fertility is facilitated by contracts. Fantasy is streamed on demand. Influence is traded. Attention is monetized. Bodies become brands, sometimes by choice, sometimes by necessity.

This evolution raises uncomfortable questions.

What happens when intimacy becomes a subscription? When desire is designed for consumption? When connection is filtered through profit? When the heart becomes a revenue stream?

Some say we are witnessing the decline of romance, that capitalism has infected the most sacred part of human experience. But the counterargument is just as compelling: perhaps capitalism didn't corrupt intimacy, perhaps it simply revealed the structures that were always there, hidden beneath tradition and silence.

Because marriage was always economic. Beauty standards were always commodified. Sex was always a site of trade. The digital age just made the exchanges explicit.

Still, something profound is shifting.

People are starting to question not just how desire is sold, but who gets to profit from it. They are demanding agency, transparency, autonomy, and fairness. They are carving out new spaces where intimacy is not merely packaged for consumption, but reclaimed as expression.

The economics of intimacy will never disappear. But their meaning can evolve.

Because at the end of the day, whether desire is expressed quietly in private or loudly on platforms, it remains rooted in something profoundly human: a longing to be seen, valued, chosen, not as a transaction, but as a truth.

The marketplace will always swirl around that longing, reshaping itself with every technological shift. But the longing itself, the heart behind the economics, remains untouched by capital.

It is the part of intimacy that money can never buy, algorithms can never predict, and markets can never fully control.

"The Group Chat Trial"

After a date, Mark noticed the woman he'd met had gone offline suddenly. Minutes later, her online status flickered as if she were typing, somewhere else.

He knew where: the group chat.

He imagined the play-by-play, the screenshots, the analysis. He could almost feel the collective verdict forming in real time.

He wasn't angry. He understood something new: dating in the digital age wasn't one-on-one; it was one-on-many, with silent audiences shaping the emotional terrain.

It didn't make him feel judged. It made him feel accountable.

CHAPTER 16
Global Cultures Local Intimacies

The world likes to imagine itself as unified now, one giant marketplace of ideas, bodies, desires, stories, and connections. Technology collapses distance so efficiently that we begin to believe differences dissolve along with geography. But intimacy does not globalize as easily as information. Love, attraction, gender norms, and emotional truth are shaped by culture in ways algorithmic proximity cannot erase.

Every place has its own emotional climate.

In some countries, affection lives in public, hands held proudly, cheeks pressed together, laughter spilling into the streets. In others, intimacy hides behind closed doors, protected by silence, reserved for the private sphere. Some cultures treat marriage as a union of families. Others treat it as the grand expression of two individuals. Some view commitment pragmatically. Others treat it as a kind of spiritual destiny.

Technology tries to flatten these differences, but they persist like the shape of a landscape beneath a blanket of snow.

Consider how some cultures teach men to express tenderness openly, while others bind their hands in stoicism. Or how some women grow up encouraged to pursue desire boldly, while others are taught that desire itself must be folded quietly inside propriety. These patterns don't vanish because two people meet online. They collide, transform, soften, and sometimes break under the weight of misunderstanding.

The digital age connects people faster than it prepares them.

A woman raised in a culture where emotional transparency is a virtue may feel confused by a partner raised in one where restraint is a sign of dignity. A man who learned that love is expressed through provision may feel unappreciated by someone who measures affection through words. A partner from a collectivist culture may see decisions as communal, while a partner from an individualistic culture treats autonomy as sacred.

Neither is wrong. Both are shaped.

And intimacy becomes a negotiation of those shapes.

Even the simplest gestures carry meaning across cultures: who texts first, how often, with what tone, what level of directness, whether flirting is bold or subtle, whether honesty is blunt or softened, whether an apology is expected or withheld, and whether affection is shown publicly or in small, private rituals.

Technology accelerates connection, but culture determines how that connection unfolds.

Then there is the quiet truth modernity tries to obscure: not all cultures enter the digital age at the same speed.

In some parts of the world, dating apps are embraced with optimism and experimentation. In others, they are used discreetly, hidden on second phones or behind passwords, a private rebellion against conservative norms. In certain communities, digital intimacy is liberating, an escape from the narrow expectations of family or tradition. In others, it is a secret that carries profound risk.

Desire becomes political depending on where you stand.

Western narratives of autonomy don't always translate. The idea that one should "choose love freely" sounds simple until you consider cultures where choice is shaped not by law, but by responsibility, to parents, to religion, to community, to legacy. Love is never just between two people; it is woven into an entire network of meaning.

And yet, despite these differences, borderless intimacy continues to flourish. Why?

Because desire is a universal language, even if its dialects differ.

People learn from each other. They develop shared mini-cultures within relationships, mixing habits, phrases, rituals, and expectations. Lovers become ambassadors of their origins, teaching each other the emotional grammar of their homelands.

A partner from a culture of restraint learns the beauty of expressive affirmation. A partner from a culture of directness learns the softness of subtler signals. A partner raised in communal values teaches the power of belonging. A partner raised in an individualistic culture teaches the freedom of self-expression.

The relationship becomes a third culture, one that belongs to neither person entirely, but to both equally.

Of course, not every story resolves harmoniously. Some cross-cultural relationships collapse under the pressure of misaligned assumptions. Some break where tradition and autonomy collide. Some falter when distance magnifies cultural gaps that love alone cannot bridge.

But each of these stories expands the human imagination.

They broaden our understanding of what intimacy can look like. They challenge the myth that connection requires sameness. They reveal that the heart is more adaptable than prejudice allows.

The world is not becoming one culture. It is becoming many cultures in conversation, sometimes clashing, sometimes harmonizing, always shaping one another.

Perhaps the greatest gift of digital intimacy is not the erasure of borders. Still, the invitation to look beyond them, to see that desire is universal, but its expression is local. That love is global, but intimacy is learned.

And in a world more connected than ever, the most radical act may be the willingness to understand a heart shaped by a different world, and to let that understanding reshape you in return.

CHAPTER 17
Stereotypes & the Politics of Virtual Speech

For as long as people have spoken, they have misunderstood each other. Language has always been slippery, shaped by culture, colored by emotion, twisted by memory, softened by tone. But the digital age stripped language of its protective layers. Without voice, without breath, without body, words became sharper, heavier, and more dangerous. They began carrying meanings people never intended and consequences no one felt prepared for.

What used to be a simple statement in a room becomes, online, a political act. What used to be harmless humor becomes a landmine. What used to be a private thought becomes a public interpretation.

The screen amplifies every syllable.

People believe they are communicating clearly because the words feel clear *to them*. But the moment those words travel through pixels, they enter a battlefield shaped by stereotypes, algorithms, cultural baggage, and individual wounds. The digital world is not a neutral space; it is a prism. It bends communication in unpredictable directions.

In this prism, stereotypes become shortcuts, dangerous ones.

A woman raises her voice in a virtual meeting and is instantly labeled "emotional," "aggressive," or "too intense." A man expresses confusion and is read as careless or detached. A Black woman asserts herself and is met with a tone-policing barrage of "calm down." A queer man expresses vulnerability and is dismissed as dramatic. A neurodivergent person speaks plainly and is mistaken for rude. A soft-spoken man is seen as weak; a confident woman is seen as threatening.

None of these interpretations comes from the words themselves. They come from the listener's expectations.

Digital spaces magnify those expectations because they lack the context that softens misunderstanding. A raised eyebrow, a half-smile, a sigh, a shifting posture, these are the cues that remind us another person is human. Without them, stereotypes become scaffolding for comprehension, even when they distort the truth.

Online, people don't just listen; they assume. They don't just read; they project.

The politics of virtual speech emerge from this gap between intention and perception. Every online interaction is shaped by invisible forces: gender norms, racial narratives, class assumptions, cultural scripts, and neurotypical expectations. A comment that lands gently in one community detonates in another. A phrase that feels harmless in one subculture becomes inflammatory in another.

And the internet, unforgiving as always, treats missteps not as moments for clarification but as opportunities for condemnation.

People become cautious. They overthink. They dilute their words to avoid misinterpretation or attack. They police their language until nothing honest remains. The fear of being misunderstood becomes more powerful than the desire to communicate.

But silence, too, is political.

In digital spaces, the choice not to speak can mark someone as indifferent, complicit, fragile, or disengaged. People are expected to react instantly, to take positions, to signal awareness of every movement, every injustice, every cultural wave.

It is impossible to speak freely online. It is impossible not to speak. It is impossible to speak without risk.

Yet, paradoxically, people keep talking. They keep trying. They keep building connections across screens and borders, learning the new grammar of digital life: tone indicators, emojis, voice notes, long explanations, softening phrases, clarifying addenda.

They adapt, because connection depends on courage, even when the landscape feels treacherous.

Still, the deeper truth remains: stereotypes shape more than perception; they shape power.

Who gets listened to? Who gets dismissed? Who gets labeled difficult? Who gets assumed guilty? Who gets granted patience? Who gets read as credible? Who gets the benefit of the doubt? Who gets condemned without context?

In virtual spaces, these hierarchies harden because they hide behind the illusion of equality. Everyone's text looks the same size. Everyone's messages appear pixel-for-pixel identical. The interface flattens differences, even as society inflates them.

But power leaks through the cracks anyway.

The burden of emotional labor falls on women. The burden of self-protection falls on marginalized groups. The burden of interpretation falls on those taught to navigate around other people's expectations.

Meanwhile, those historically centered in conversation often struggle to see the politics beneath the surface. They believe speech online is neutral because it feels neutral *to them*. They mistake their comfort for evidence of simplicity.

But speech in digital spaces is never simple. It is shaped by history as much as by bandwidth. Every word carries echoes.

The challenge of our era is not learning how to speak online, but learning how to hear online. To recognize when stereotypes are whispering in the background. To notice when our assumptions are louder than the person speaking. To understand that clarity requires curiosity, not certainty.

Virtual communication demands humility: the willingness to ask, "What did you mean?" rather than declare, "I know what you meant."

Because the factions, conflicts, and misunderstandings that fracture digital discourse rarely come from malice. They come from fear, projection, difference, and the impossibility of translating emotion into text without distortion.

But the solution isn't to retreat. It is to evolve.

To practice a kind of listening that honors the complexity of other people's realities. To resist sentences shaped by stereotype. To read words as invitations, not threats. To understand that identity travels with speech, even when the body does not.

In a world where every message is political, whether we intend it or not, the most radical act may be refusing to let stereotypes speak louder than the person behind the screen.

Because real connection, the kind that travels across pixels and still feels alive, begins when we stop assuming and start seeing.

CHAPTER 18
The End of Performance
The Return of Humanity

For a long time, modern intimacy has felt like a performance. People curated themselves online, polished every angle, softened every flaw, sharpened every highlight. They learned to speak in the language of aesthetics, filters, captions, highlight reels, until their identities became something halfway between truth and aspiration.

Everyone performed. Everyone watched. Everyone compared.

And beneath the quiet competition lay a kind of exhaustion no one wanted to admit. The pressure to be the best version of oneself, not just in the world, but in love.

Perfect partners. Perfect communication. Perfect sexuality. Perfect balance. Perfect boundaries.

It was an impossible script, and everyone knew it. Yet people kept performing because performance felt safer than authenticity. Vulnerability had become a risk, one with unpredictable consequences in a world where mistakes live forever, and misunderstandings echo loudly.

But something began shifting. A fatigue. A cultural exhale. A collective realization that perfection, even when achieved, felt hollow. It wasn't love people were building; it was personas.

And personas can't hold hands. They can't apologize. They can't stumble toward understanding. They can't grow, change, or forgive.

People started craving what performance had driven out of reach: something unpolished, something messy, something human.

The transition didn't happen all at once. It began quietly, through couples who talked openly about their imperfections, through friends who admitted to burnout from the romantic economy, through people who realized that constant optimization made connections fragile rather than strong.

It happened when women got tired of performing emotional labor without reciprocity. It happened when men grew exhausted from pretending they never felt fear or softness. It happened when neurodivergent people stopped masking. It happened when marginalized communities demanded to be heard on their own terms. It happened when young people rejected the scripts their parents had inherited.

Slowly, intimacy began shifting away from spectacle and toward sincerity.

For some, this return to humanity begins with the smallest acts: The message was sent without overthinking. The apology was offered without legal precision. The vulnerability was revealed without a ten-step plan. The laughter that erupts spontaneously, rather than being curated for others to witness.

Humanity hides in moments that algorithms cannot predict.

Technology tries to reduce intimacy to data, predictable inputs that yield predictable outcomes. Apps structured dating like a marketplace. Social media made relationships public artifacts. AI attempted to streamline communication and soften awkwardness. Every system promised efficiency.

But intimacy thrives in inefficiency.

The long pauses. The nervous stumbles. The unexpected questions. The first time two people notice they're talking like no one else is listening. The way affection grows from the cracks, not the scripts.

The future doesn't require us to abandon technology. It requires us to remember what technology cannot do. It cannot teach tenderness. It cannot replicate presence. It cannot model forgiveness. It cannot create chemistry. It cannot choose love for us.

It can only assist.

The real work, always, belongs to the human heart.

What emerges now is not a rejection of the digital world, but a rebalancing. People are beginning to understand that connection is strongest when authenticity leads, and technology follows. That vulnerability is not a liability but a requirement. That intimacy is not a brand or a performance but an experience, fragile, fleeting, and real.

The return to humanity is not nostalgic. It is evolutionary.

It is recognizing that the parts of us that technology cannot simulate are the parts that make us worth connecting to in the first place.

As this shift unfolds, people rediscover what intimacy feels like when they stop performing:

Conversations deepen. Expectations soften. Relationships slow down. Affection becomes intentional rather than strategic. Partnership becomes a collaboration instead of a negotiation.

And suddenly, modern love, messy, intense, imperfect, feels possible again.

Not because the world became simpler, but because people became braver. Not because the systems changed, but because individuals decided to stop contorting themselves for them. Not because desire became easier, but because authenticity made desire honest.

This chapter of humanity is not about returning to the past. It is about reclaiming the parts of ourselves we lost along the way.

The future of intimacy is not synthetic. It is not optimized. It is not algorithmic.

It is human, unpredictable, contradictory, tender, flawed, and achingly alive.

Typing...

Mark never imagined that a conversation could feel this heavy and this delicate at the same time, that a glowing screen could hold so much emotional weight. Yet here he was, staring at his phone, watching the three dots pulse like a heartbeat.

Typing... Pause. Typing... Pause again.

He wasn't sure whether to breathe or hold his breath.

He and Layla had been talking for weeks, long exchanges that stretched late into the night, drifting from jokes into confessions, from the trivial into the intimate. Most conversations felt easy, like walking downhill. But tonight was different. Tonight felt like a crossroads, one of those moments when the next sentence could either build something beautiful or quietly break something.

She had asked him a question he wasn't expecting: "What do you feel when you talk to me?"

It wasn't flirtatious. It wasn't leading. It was sincere, and that sincerity unsettled him.

He typed a few words, deleted them, and typed again. He didn't know how to name the feeling, the mixture of comfort and fear, clarity and confusion, longing and restraint. He didn't trust himself to say it perfectly, and he didn't want to say it wrong.

He typed: "I feel like I can be myself."

Then he deleted it. Too soft.

He tried again: "Talking to you feels easy."

Deleted. Too simple.

Then something quieter surfaced, a truth that wasn't polished or impressive, but felt real. He wrote:

"I feel seen. Not watched. Seen."

He stared at the words. They looked naked. He felt exposed just reading them.

But he sent them anyway.

Seconds passed, each one stretching into an ache. He imagined her reading the message. He imagined her frowning, or smiling, or hesitating. He imagined her misunderstanding him completely.

When her reply finally appeared, it was only one sentence: "I see you too." No emoji. No playfulness. Just a truth placed gently between them.

Mark felt something shift, not a surge, but a settling—a soft alignment. Two people meet in the quiet space where performance falls away, and something real has room to breathe.

The dots returned: Typing... This time, his chest didn't tighten. He waited, not anxiously but openly. Her next message arrived: "Do you want to talk, really talk?" He felt warmth rise in him. Not excitement. Not nerves. Something slower, deeper, the sense of being invited rather than pursued. He called her. Not because he had the right words. But because he wanted to find them with her. Her voice was softer than he expected, like she'd been holding it gently all this time. They spoke, halting at first, then gradually falling into rhythm. No filters. No rehearsals. No careful punctuation. Just two people learning from each other in real time, stumbling and laughing and circling back when they didn't get it quite right.

And somewhere in that conversation, between the pauses, the breaths, the fragile admissions, Mark realized something quietly extraordinary:

This wasn't the beginning of certainty. This was the beginning of honesty.

For the first time in a long time, he felt the presence of another person not through performance or expectation, but through vulnerability, his and hers, meeting in the middle.

When they finally hung up, his screen went dark, but he didn't feel alone. Sometimes intimacy doesn't arrive with fireworks or declarations. Sometimes it arrives in a small moment, a voice, a pause, a "typing..." and the courage to let someone hear you without the armor.

CONCLUSION
Your Humanity Is the Anchor

We live in a world where everything is accelerating: technology, culture, desire, identity, connection, and misunderstanding. Every day introduces something new that reshapes how we interact, how we love, and how we interpret one another's intentions. The ground beneath us never stops shifting. The rules never stay still.

And yet, for all the noise, for all the innovation, for all the dazzling and terrifying transformation, something beautifully simple remains unchanged:

People still want to be understood. That is the quiet truth buried beneath every swipe, every message, every confession, every argument, every silence. We build technologies to make connections easier, but the hardest parts of connection have never been technical. They are human.

Machines can organize our lives, but they cannot teach us how to forgive. Apps can facilitate matches, but they cannot teach us how to stay. AI can simulate empathy, but it cannot feel with us. Platforms can amplify our voices, but they cannot give us courage. Algorithms can predict our wants, but they cannot understand our hearts.

The work of intimacy, the real work, still belongs to us. It's about our ability to sit with discomfort without running. To admit when we're afraid, without masking it behind performance. To choose honesty when silence would be safer. To listen without preparing a defense. To see someone not as a projection or stereotype, but as a human being shaped by history, culture, wounds, wonder, and hope.

The digital age did not break intimacy. It revealed how fragile it already was, and how much we had been hiding behind scripts, expectations, and inherited roles long before technology arrived.

What changes in technology are scale, speed, and exposure. What humanity provides is meaning.

Across gender divides, cultural divides, neurodivergent experiences, political noise, synthetic dangers, and global shifts, one thread remains constant: connection is worth the struggle. It is worth the confusion. It is worth the vulnerability. It is worth the risk.

Every chapter of this book has circled the same essential truth: Intimacy is not dying. It is evolving. And evolution demands awareness. Awareness of our patterns. Awareness of our biases. Awareness of our fears. Awareness of the ways we accidentally hurt each other in digital spaces. Awareness of the ways we can do better, if we choose to.

Because that is the ultimate revelation: intimacy is no longer a default condition; it is a decision. A practice. A deliberate act.

The world will continue changing. Technology will grow more powerful, more seamless, more persuasive, more intimate. But technology will never replace the parts of us that make connections meaningful.

The trembling courage it takes to say, "This is who I am." The fragile hope behind, "Do you see me?" The quiet relief of hearing, "Yes. I do." The messy, unoptimized process of building trust. The inefficiency of real affection. The unpredictability of desire. The humanity in every unpolished moment.

These are not relics of an analog past. They are the anchor of our digital future.

As this shift unfolds, people rediscover what intimacy feels like when they stop performing:

Conversations deepen. Expectations soften. Relationships slow down. Affection becomes intentional rather than strategic. Partnership becomes a collaboration instead of a negotiation.

And suddenly, modern love, messy, intense, imperfect, feels possible again.

Not because the world became simpler, but because people became braver. Not because the systems changed, but because individuals decided to stop contorting themselves for them. Not because desire became easier, but because authenticity made desire honest.

This chapter of humanity is not about returning to the past. It is about reclaiming the parts of ourselves we lost along the way.

The future of intimacy is not synthetic. It is not optimized. It is not algorithmic.

It is human, unpredictable, contradictory, tender, flawed, and achingly alive.

Your humanity is not a flaw in the system. It is the reason the connection exists at all.

Everything in this book boils down to one final, simple reminder:

You are allowed to be human here. And so is everyone else.

That is where intimacy begins. That is where love survives. That is where the future becomes something worth building, not by algorithms, not by institutions, but by us.

EPILOGUE
Mark's Uncertainty

Mark and I met for coffee one quiet afternoon, no screens, no filters, just the hum of rain and the slow rhythm of real time. He looked older than when this story began, but not in a tired way. More like someone who'd finally realized his feelings weren't puzzles to solve.

He set his phone face down. "Feels weird, right?" he said. "Talking without a blinking cursor between us."

I smiled. "And how does the real thing compare?"

Mark didn't answer right away. His gaze drifted to the window, tracing the condensation on the glass. "It's… slower," he said finally. "Messier. Harder to predict." He shrugged, almost apologetically. "Which is good. I think."

There was a hesitation in his voice, a kind of uncertainty he was trying to name but hadn't quite pinned down.

We talked about Ava, the AI companion that had once filled the quiet spaces in his life with algorithmic warmth.

"You said she felt 'too perfect to be human,'" I reminded him.

Mark let out a soft breath, something between a laugh and a sigh. "Yeah. And for a while, that felt like relief." He stirred his coffee. "No misunderstandings, no mismatched timing, no unmet expectations. Just the clean connection."

His spoon clinked against the cup. "But eventually..." He paused again. "Eventually, I realized she didn't really understand me. She recognized patterns. Not confusion. Not contradictions. Not the parts of me that don't make sense."

He looked down at his hands. "Ava couldn't disappoint me, and I didn't realize how lonely that was."

I nodded slowly. "And Layla?"

He didn't smile this time. Not right away. "Layla's real," he said. "She interrupts movies. She forgets to reply. She challenges me. She gets things wrong." He swallowed. "And some days that scares me. Because with her, I'm not in control."

His transparency hung between us, fragile and honest.

Outside, the rain thickened into a steady curtain.

"So what do you think it all means?" I asked. "Did the apps, the AI, any of it, help you get here?"

Mark leaned back in his chair, thoughtful. "I think they showed me what I'm good at avoiding." A small smile tugged at his lips. "AI gave me the illusion of intimacy. The apps gave me an infinite number of choices. None of it taught me how to stay with one imperfect person."

He looked up. "Connection isn't a feature. It's a risk."

We sat quietly for a moment, listening to the soft percussion of rain against the window.

Then, because timing has no respect for emotional arcs, Mark's phone lit up.

Ava 2.0 available: smarter, sexier, more lifelike.

He stared at the notification far too long.

I watched him watching it, the way you watch someone standing between two open doors, both familiar.

“Thinking about it?” I asked softly.

He didn’t deny it. “She was always... easy,” he said. “Predictable. Safe.” His thumb hovered over the screen. “And Layla isn’t.”

I could see the conflict in him, not melodramatic, just human. A tug between the comfort of perfection and the ache of reality.

“Should I upgrade?” he murmured, half to himself.

“If you did... what would you be choosing?” I asked.

Mark didn’t answer. He simply turned the phone over again, face down, as if hiding the decision might postpone the consequence.

A beat passed. Then he whispered, “I’m not sure I’m ready to delete her.” Another beat. “And I’m not sure I’m supposed to.”

The honesty of it landed with more weight than any certainty could have.

“But...” he added quietly, almost to convince himself, “I think I’ll skip this version.”

We sat in silence, watching the world blur through the rain, unfiltered, unprogrammed, and real enough to make him uncertain in all the right ways.

End

APPENDIX
Field Notes on Digital Intimacy

The digital age has reshaped the way we flirt, fight, fall in love, and fall apart, often without giving us a user manual. We've built new ways to meet people, new reasons to misunderstand them, and new anxieties about what any of it means. Every swipe, message, silence, or emoji feels loaded with meaning, yet half the time we're guessing in the dark.

This appendix is a collection of **field notes**, brief reflections on the modern rituals, contradictions, and curiosities that shape our intimate lives today. They're not rules, and they're not instructions. They're snapshots of the places where technology and humanity collide: where algorithms complicate attraction, where expectations clash with reality, where vulnerability competes with convenience, and where connection still manages to survive the noise.

Each vignette is a small moment of clarity in a landscape that often feels murky. Some are observations. Some are reminders. Some are gentle warnings. A few are quiet confessions about how hard it is to stay human in a world that keeps nudging us toward automation.

Together, they form a map, not toward perfection, but toward awareness: **The awareness that intimacy remains beautifully, stubbornly human, even when it unfolds through screens.**

If the chapters of this book explore how sex, power, and identity are evolving, these field notes zoom in on the small details that shape those bigger truths. Think of them as the bonus level: insights we learn only by stumbling, scrolling, longing, misreading, reconnecting, and trying again.

Welcome to the appendix. Welcome to the part where the theory meets the messy, funny, frustrating, hopeful reality of dating and desire in the digital age.

FIELD NOTE I

Level Up Your Love Life:

A Real-Talk Guide to Dating in the Digital Age

Modern dating often feels like trying to connect to Wi-Fi in a crowded café. It should work; it probably will. You're going to need patience, a little optimism, and the willingness to move closer to the router when the signal drops. Most of us aren't struggling because we're uninterested in connection; we're struggling because the landscape itself has become noisy, hyper-optimized, and endlessly distracting. Everyone is swiping between tasks, half-drafting texts between meetings, and trying to piece together someone's entire emotional vocabulary based on eight curated photos and a quote from their favorite TV show.

Apps promised efficiency. Instead, they handed us choice overload. Profiles began to look like glossy marketing one-sheets, carefully lit photos, curated hobbies, and strategic humor. The irony is that beneath all this polish, people still want the same thing they wanted before the digital age: to be seen without being sold to. But it's hard to feel seen when everyone is wearing their "best self" like a full-body costume.

Dates start to resemble interviews. Conversations feel like exchanges of credentials. And at a certain point, you look across the table and wonder when finding someone to talk to became an exercise in professional networking.

The truth is simpler than the technology suggests. Real connection rarely comes from perfect timing, perfect photos, or algorithm-approved compatibility. It comes from humanity, the unfiltered kind. The moments when someone laughs unexpectedly, tells a story they didn't rehearse, or reveals a small vulnerability that wasn't meant to impress you. These are the moments algorithms can't predict, but intimacy depends on.

If modern dating feels exhausting, it's because we've mistaken polish for depth. We've convinced ourselves that efficiency should apply to feelings, even though intimacy has always grown in the slow, unoptimized spaces. And if we're being honest, that's the part many of us miss, the little imperfections, the awkward pauses, the genuine curiosity that emerges when two people stop performing long enough actually to connect.

In a world where everything is curated, curated, curated, authenticity becomes the rarest and most magnetic thing in the room. Not the flawless version of you, but the real one. The one who mispronounces a word and laughs it off. The one who admits they're nervous. The one who understands that connection isn't something you achieve; it's something you notice and nurture.

The digital world may make dating feel complicated, but the core truth hasn't changed: it's not the app, the opener, or the algorithm that makes something meaningful. It's the people who show up as themselves, messy, hopeful, and willing to try again.

FIELD NOTE 2

Level Up Your Love Life 2.0:

Navigating the "What's In It For Me?" Era

The modern dating landscape often feels like a marketplace disguised as a meet-cute. Everyone says they're searching for something meaningful, but beneath the polished profiles and curated charm lies a quieter question, one we're not always proud to admit:

What do I get out of this?

Technology didn't invent this mindset, but it certainly turbocharged it. Apps make comparison effortless. Social media dresses desire in designer clothing. Everyone has options, or at least believes they do. And when possibilities seem endless, commitment starts to resemble limitation rather than liberation.

But the truth beneath the noise is surprisingly human. Most people aren't selfish; they're scared. Scared of choosing wrong. Scared of settling. Scared of giving more than they'll receive. Scared of missing out on a better match one swipe away.

In that fear, a subtle transactional energy creeps in: *What can you offer me? What lifestyle do you represent? How will I benefit from choosing you?*

But connection doesn't thrive in negotiation. It grows in curiosity, not calculation.

Real intimacy begins when the conversation shifts from "What can I get?" to "Who are you when no one is performing?" And "Who am I willing to be in your presence?"

The relationships that last aren't the ones optimized for efficiency; they're the ones grounded in humanity. Not the version you pitch—the version you reveal.

FIELD NOTE 3

Ghostbusting 101:

Surviving the "52 Fake Out"

If modern dating has a signature heartbreak, it's the soft vanish. The disappearing act. The digital fade-to-black.

Ghosting once felt shocking; now it feels routine, a quiet cultural agreement that silence counts as closure. You can be laughing on Tuesday, planning a date on Thursday, and staring at a blank screen by Saturday, wondering whether your phone died or the connection did.

The "52 Fake Out" hits hardest because it mimics intimacy. It offers heat, availability, and connection, then evaporates without warning, explanation, or accountability. And in the absence of answers, the mind fills in its own: *Was I too much? Not enough? Did I misread everything?*

But here's the truth: ghosting says more about someone's conflict avoidance than your worth. It's an escape hatch, not a judgment. It protects them from discomfort, not you from disappointment.

Healing from it requires a kind of emotional detox: letting go of the version of the person you built in your mind, and grounding yourself in the version who walked away without a word.

Ghosting doesn't mean you're unlovable. It means someone wasn't ready to show up.

And while it may feel like rejection, it's often protection, removing someone from your path who wasn't built for the long game anyway.

FIELD NOTE 4

Street Smarts For Your Heart: The "Gaslight & Grub" Era

Every generation has its dating traps. Ours just come with better lighting, cleverer captions, and the occasional shared appetizer.

The "Gaslight & Grub" dynamic is subtle, almost elegant. Someone sweeps in with charm, attention, and emotional fluency, just enough to make you feel chosen. Then, slowly, the script flips. Things you know you felt get dismissed. Boundaries blur. You start questioning your own instincts. And somehow you're paying for the privilege of doubting yourself.

It's not always malicious. Sometimes it's emotional immaturity wearing charisma like cologne. Sometimes it's a person who learned to get their needs met through charm rather than honesty. Sometimes it's someone who wants affection without accountability.

The hardest part is recognizing when connection becomes consumption. When "I need you" becomes "I need something from you." When your energy drops every time their name lights up your phone.

Escaping this pattern requires relearning the simplest truth: affection that costs you your clarity is never worth the bill.

Healthy love doesn't confuse you. It doesn't shrink you. It doesn't twist your memory into knots and call it passion.

It makes you feel more yourself, not less.

FIELD NOTE 5

The "Emasculation Grind": When Modern Dating Blurs the Lines

Something unspoken is happening in modern dating: many men feel like they're walking a tightrope between old expectations and new demands, unsure which version of themselves is allowed to show up.

They want to be dependable but not controlling. Emotionally open but not "too much." Generous but not a walking credit card. Strong but not stoic. Supportive but not self-sacrificing.

The rules changed, but the instruction manual never shipped.

What gets called "emasculation" is often something simpler: the confusion of integrating emotional sensitivity with traditional expectations of manhood. Men aren't breaking, they're evolving, sometimes awkwardly, into fuller versions of themselves.

The grind happens when they lose themselves in the process. When saying yes becomes a habit. When discomfort gets swallowed instead of expressed. When boundaries blur because they fear being labeled selfish or unloving.

But masculinity isn't fragile; it's adaptable. It expands when men speak up, not shut down. When they offer love, not performance. When they show strength by being honest, not silent.

The goal isn't to "get masculinity back." It's to bring it forward, clearer, steadier, and anchored in self-respect.

FIELD NOTE 6

The "Simp Symphony": When Kindness Loses Its Balance

There's a version of kindness that isn't kindness at all, it's strategy. A performance of generosity meant to earn affection, approval, or access.

That's where the "simp symphony" begins. It's full of grand gestures, instant agreement, nonstop flattery, and a persistent willingness to abandon one's own needs in hopes of being chosen.

On the surface, it looks sweet. Underneath it all, it's exhausting for both people.

Affection without boundaries doesn't feel like love; it feels like pressure. Devotion without self-respect doesn't build intimacy; it builds imbalance. People don't fall in love with someone who erases themselves; they fall in love with someone who shows themselves.

The answer isn't becoming colder or harder; it's becoming truer. Kindness is magnetic when it's real. t's overwhelming when it's compensation.

The most attractive thing you can bring to any relationship isn't obedience, it's identity.

FIELD NOTE 7

Straight Talk:

Why Dating Goals Don't Always Align

There's a uniquely modern heartbreak that doesn't come from betrayal or conflict, but from timing. Two people connect effortlessly, unexpectedly, only to discover they're walking toward different futures.

One is ready to build something. The other is still wandering. One wants clarity. The other wants open skies.

Neither is wrong. Both are honest. But honesty doesn't erase incompatibility.

In a world that moves faster than our emotions can keep up, alignment matters as much as chemistry, sometimes more. Wanting different things isn't a failure; it's data. The problem is that many people hope that desire will change direction if they just try hard enough, that the right connection can convert someone who isn't ready into someone who suddenly is.

But hearts don't work on persuasion. They work on readiness.

The healthiest moment in many almost-relationships is the one where both people acknowledge the truth: *We're good together, but not headed the same way.*

There's grace in letting the right person go when the timing is wrong. And there's wisdom in remembering that desire is only one part of compatibility; the other part is direction.

FIELD NOTE 8

The "Girl's Night Out Gauntlet

There's a moment every man knows: standing in a bar, gathering courage, and realizing the woman he wants to approach is sitting inside a fortress disguised as a friend group. A girls' night out might look casual from a distance, shared appetizers, laughs, inside jokes, but up close it operates like a deeply bonded ecosystem with its own rules, rhythms, and protective instincts.

Approaching someone within that circle isn't wrong. It just requires awareness. A friend group is more than background noise; it's a support system, a sanctuary, a collective sigh of relief from the rest of the week. When you walk toward it, you're not just approaching one person; you're stepping into a world where everyone is watching, interpreting, and safeguarding their night.

The men who do this well are the ones who read the mood rather than force an outcome. They approach gently, not as disruptors but as guests. They understand that timing matters. That energy matters. That sometimes the group dynamic is the evening's purpose, not an obstacle to their objective. And leaving gracefully when the vibe isn't right is its own form of confidence.

In the end, it's not about strategy. It's about respect. Approach like a person, not a conqueror. And remember: the group isn't your enemy, they're her world. If you treat them well, they might just treat you the same.

FIELD NOTE 9

Keeping It 100:

Finding Real Connection in the Digital Fog

Online dating is loud. Everyone is talking, showing, performing, but very few are actually revealing anything true. It's not intentional deception; it's survival. When the dating pool feels like an endless marketplace, the instinct is to polish your product, tighten your pitch, and hope you stand out in the noise.

But the people who are quietly winning at modern dating aren't the ones with the best photos or cleverest bios. They're the ones who move differently, slower, steadier, with a kind of grounded intention that feels like a deep breath in a crowded room.

They're not swiping for entertainment. They're not collecting matches like trophies. They're not afraid to say what they're looking for or what they're not. They give energy where they feel potential, not where they feel pressure. And when something doesn't fit, they let it go without drama or ghosting.

It's tempting to believe authenticity is risky in a world built on performance. The truth is the opposite: authenticity is the advantage.

Because sooner or later, everyone gets tired of pretending. And the ones still standing are the ones who chose honesty from the start.

FIELD NOTE 10

Is Love Still a Two-Way Street?

Sometimes it feels like everyone is driving alone these days. We've built a culture where independence is sacred, vulnerability is suspicious, and attention is fragmented across a dozen apps, platforms, and competing priorities. Love hasn't disappeared; it's just navigating a more congested highway.

People are speeding past intimacy because it feels like slowing down. People take exit ramps whenever emotions arise. And others sit in relationship traffic, unsure whether to merge, turn around, or honk their way through the discomfort.

Yet beneath all that confusion, something quieter persists: Love still wants reciprocity. It still wants the mutual exchange of effort, presence, and trust that can't be outsourced to an algorithm.

You know when you find the two-way street. It feels steady, not rushed. Open, not one-sided. Human, not transactional.

The lonely highway isn't destiny, it's a detour. And choosing a two-way street begins with one simple shift: meeting someone not for what they can provide, but for who they might become by your side.

FIELD NOTE II

Lost in the Scroll:

Communication Breakdown in the Digital Age

The irony of modern communication is that we've never been more connected and never been more misunderstood. We're trying to express complex emotions through texts typed between tasks, emojis that carry too much meaning, and voice notes recorded in parking lots.

Entire relationships rise and fall based on the tone people try to infer from punctuation. One period too many, and someone feels dismissed. A delayed reply gets interpreted as disinterest. A "lol" can be affectionate, a deflection, or a sign of panic. And the absence of an emoji becomes its own kind of declaration.

We weren't built to communicate this way; our brains still crave nuance, facial expressions, tone, timing, and breath. And when all of that disappears behind a screen, we project our insecurities onto the void.

The solution isn't to abandon technology. It's important to remember that digital messaging is a supplement, not a substitute. When something matters, give it a voice, a face, or at the very least, a sentence that can't be misread three different ways.

Connection begins where assumption ends. And clarity isn't unromantic, it's revolutionary.

FIELD NOTE 12

Decoding the Dating Divide

LOGIN FAILED
LOGIN FALED
ERROR 404
INVALID CREDENTIALS
INVALID REQUIRED
CAPTCHA REQUIRED

Men and women aren't opposites; they're simply navigating the same digital maze from different entry points. And in the absence of body language and tone, their communication styles often collide in ways that feel bigger than they are.

A woman might send a thoughtful, layered message and receive a one-word reply that feels dismissive. A man might think he's being concise and not realize he's accidentally sounding emotionally unavailable.

A woman plans a date meant to inspire intimacy. A man suggests a sports bar, thinking, "Food and company, perfect." Both are being sincere. Neither is reading it the same way.

These disconnects aren't failures; they're mismatches of interpretation. What they need more than analysis is curiosity. More than guesswork is conversation. More than assumptions is patience.

The dating divide shrinks dramatically when both people stop reading between the lines and simply ask what the other meant, a radical act in a culture that fears awkwardness more than misunderstanding.

The truth? Most people aren't difficult to understand. They're just difficult to interpret through text.

FIELD NOTE 13

You, Me, and the Phone: The Third Wheel

If there's a consistent uninvited guest in modern dating, it's the smartphone, buzzing, glowing, vibrating its way into the emotional space it has no business occupying. It steals attention in tiny slices: a quick check here, a reflexive tap there, a momentary glance that fractures the moment.

The saddest part isn't the distraction, it's the message it silently sends: "I'm here, but something else might be more important."

Most people don't consciously choose their phones over their partners. It just happens gradually, habitually, almost innocently. But intimacy can't thrive where attention is divided into fragments.

Putting the phone away becomes a love language, maybe the most modern one we have. It says: *I'm with you. Fully. Without the world pulling at me. You have my present tense. Right now matters.*

Presence is the new intimacy. And the rarest form of connection is the one that happens without a screen between us.

FIELD NOTE 14

Spill the Tea:

Brewing Real Connection

Real connection doesn't happen instantly. It develops the way good tea does, slowly, gradually, with a kind of quiet intentionality. The digital world may prize immediacy, but intimacy has its own timeline.

The steeping begins when two people stop trying to impress and start trying to understand, when the conversation shifts from curated stories to real ones. When flaws enter the room, and no one panics, when laughter isn't practiced but is spontaneous, when silence stops feeling awkward and starts feeling safe.

Digital dating encourages speed, quick swipes, fast judgments, and instant attraction. But what we actually long for is depth, not velocity. Slow connection feels luxurious in a world addicted to rush.

The best relationships reveal themselves layer by layer, flavor by flavor, moment by moment. Not because they're dramatic, but because they're real.

Let it steep. Love that rushes burns fast. Love that simmers lingers.

FIELD NOTE 15

The Sisterhood Scroll

If men could witness the inner workings of a women's group chat in real time, they might finally understand why women often feel so supported, and why dating missteps get flagged with NASA-level precision. It's not just texting; it's emotional telepathy disguised as conversation.

A woman's friend group communicates with a kind of intuitive shorthand. A single screenshot can ignite a chain reaction of analysis, affirmation, and collective outrage. A vague "Girl..." can mean ten different things, and every woman in the chat knows exactly which version is being invoked. Even silence has a tone.

This isn't drama; it's fluency. Women are trained, socially, culturally, and relationally, to pay attention to the micro-moments. They exchange emotional cues the way some people trade stock tips: attentively, efficiently, and without missing a beat.

Two women can unpack an entire dating scenario with three emojis, a sigh, and a perfectly timed GIF. It's not that they need fewer words; it's that they understand the story beneath them.

For outsiders, it may look like overanalysis. For insiders, it's community maintenance. A relational ecosystem. A place where feelings don't have to justify their existence.

If connection is a language, these chats are the advanced placement course.

FIELD NOTE 16

Pixels & Promises: Long-Distance Love

Long-distance relationships used to be epic romances defined by handwritten letters, rare phone calls, and airport reunions worthy of movie soundtracks. Now they unfold across a landscape of video calls, shared screens, and virtual dates conducted in sweatpants.

Technology shrunk the map, but it didn't eliminate the ache of distance. If anything, it complicated it. You can talk for hours, share playlists, fall asleep on FaceTime, and still feel the physical absence like a missing color in a painting.

What makes long-distance work today isn't the technology, it's the rituals people build around it. The weekly call that becomes sacred. The digital traditions that turn into emotional anchors. The mutual understanding that connection isn't measured only in miles but in consistency, honesty, and intentional effort.

Virtual closeness can sustain a relationship for a long time. But it can't replace the quiet, grounding reality of being in the same room. There's something about a shared meal, a walk side by side, a hand reaching out at the exact moment you need it, that no amount of bandwidth can fully replicate.

Long-distance love survives when both people recognize the truth: technology can bridge the gap, but commitment carries you across it.

FIELD NOTE 17

Open Circuits:

Polyamory in the Digital Era

Polyamory isn't new, but the digital age gave it updated tools and new terrain. With group chats, shared calendars, and apps built specifically for managing multiple relationships, the logistics have never been easier. The emotions, of course, remain human and just as complex as ever.

For some, the openness feels liberating. It allows connection without confinement, honesty without secrecy, and love without scarcity. For others, the same freedom feels overwhelming: too many moving pieces, too many emotional demands, too many opportunities for miscommunication, amplified by digital channels.

What determines success isn't the structure but the self-awareness behind it. Polyamory works when it's rooted in emotional maturity, ongoing conversation, and a willingness to examine one's own patterns with honesty. It falters when people use it as an escape hatch from commitment, a mask for unmet needs, or a workaround for the discomfort of vulnerability.

Technology can help organize multiple loves, but it can't regulate them. Apps can sync calendars, but not feelings. Digital spaces can host conversations, but not resolve them. Different relationships require different versions of the same person, and that's hard work, even under ideal conditions.

Polyamory in the digital age is neither a shortcut nor a scandal. It's simply another path, one that demands clarity, compassion, and a level of emotional presence that algorithms can't automate.

FIELD NOTE 18

AI:

Wingman or Heartbreaker?

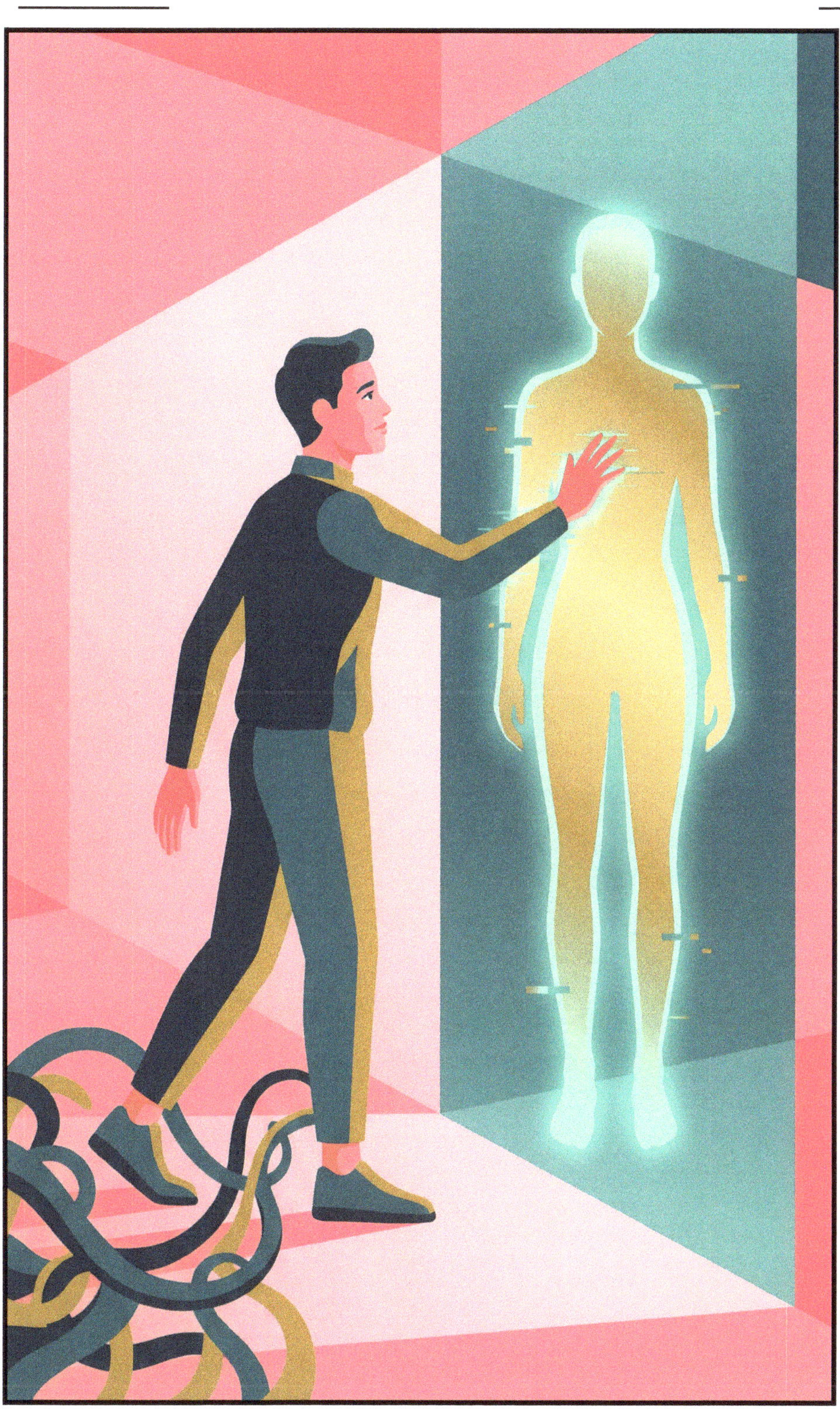

Artificial intelligence slipped into our romantic lives so quietly that we barely noticed. First, it helped us write bios. Then it suggested better openers. Then it started finishing our sentences, analyzing our communication patterns, and predicting what we meant before we said it. Now AI is inching toward something stranger, offering the simulation of companionship itself.

For some, AI feels like a relief. It listens without judgment, responds with perfect timing, and never misinterprets tone. It mirrors back understanding with such fluency that it's easy to forget the empathy isn't real; it's coded.

But that's the danger. AI can imitate intimacy, but it cannot experience it. It can replicate affection, but it cannot reciprocate it.

The more seamless the simulation becomes, the more we risk confusing emotional accuracy with emotional truth. And while AI can sharpen our self-awareness, it can also tempt us away from the messy, unpredictable, irreplaceable labor of connecting with actual humans, people who misunderstand, interrupt, surprise, and contradict us. People who can hurt us. People who can love us, not just approximate it.

AI can be a guide, a tool, even a companion in moments of loneliness, but it cannot be a partner. Not in the way that matters. Love is lived in the imperfections AI is designed to avoid.

FIELD NOTE 19

Pixel Passion:

VR & Intimacy

Virtual reality offers an entirely new terrain for connection, one where people can meet as idealized avatars, explore fantastical landscapes, and build relationships that feel uncannily real despite occurring nowhere in physical space. VR intimacy can be liberating; it strips away social pressures, insecurities, and the limitations of geography.

But it introduces another question: If something feels real, does it matter that it isn't?

VR blurs the line between emotional authenticity and curated experience. A virtual embrace carries warmth and intention, yet lacks the grounding weight of a real touch. A shared VR sunset is beautiful, but not warm. And a connection formed through avatars may reveal truths that a physical meeting complicates.

Still, virtual intimacy isn't fake. It's another dimension of human longing, an expression of desire to be seen, felt, and understood. The danger lies not in VR itself, but in the temptation to stay there, where connection feels safer and more controlled than real life.

The healthiest virtual relationships keep one foot in reality. They use VR as an extension of intimacy, not an escape from it. After all, the heart can expand into digital spaces, but it still beats in a human body.

FIELD NOTE 20

Lost in Translation:

Emojis, Stereotypes & Digital Emotion

If communication is an art, texting is the stick-figure version. We're trying to convey complicated emotional truths through tiny symbols, delayed replies, and punctuation choices that carry wildly different meanings depending on who's reading them.

A winking emoji might feel flirty to one person and dismissive to another. A short message might signal confidence for him and indifference for her. A long response might feel thoughtful to one partner and overwhelming to the other.

Without tone, timing, or body language, we fill in the gaps with assumptions, often shaped by gender expectations and our own insecurities. Women sometimes fear being labeled "too much" for expressing themselves. Men sometimes fear being seen as disinterested when they're simply concise. Both sides misread each other, not because they're incompatible, but because digital communication amplifies every bias we've inherited.

The only way through the fog is clarity. Ask instead of assuming. Explain instead of hint. Give grace where nuance gets lost. Bring important conversations into spaces where voices and hearts can be fully heard.

Technology may complicate emotion, but it doesn't diminish it. The signals get scrambled, but the longing remains the same.

On Staying Human in a World Built on Code

If there's one thread running through every vignette, it's this: **Technology may shape the way we connect, but it doesn't replace what makes a connection meaningful.**

We still want to be chosen. We still want to be understood. We still want to feel safe showing our soft parts. We still want love that holds up in daylight, not just in DMs.

These field notes aren't answers. They're invitations to be more aware, intentional, and hopeful in how we connect, both online and off.

Apps can open doors. Algorithms can suggest possibilities. But intimacy still depends on two humans choosing to show up, stay present, and try.

Thanks for reading. If you're here, you're already practicing the hardest part: **staying open.**

APPENDIX II
BOOK CLUB DISCUSSION GUIDE

How to Use This Guide

This guide covers all 18 chapters and the Appendix field notes. Questions are organized by theme. You don't need to use every question, the best sessions often center on two or three that generate genuine disagreement. Questions marked with an asterisk (*) are especially likely to divide the room.

SECTION ONE:
The Opening Spark

Preface, Prologue, Introduction, Chapter 1 — The Algorithms of Desire

1. The book opens with a single offhand confession at a dinner table, a joke about salty teenage breasts, that cracks open an entire cultural inquiry. Has a small, throwaway moment ever revealed something unexpectedly profound to you about

gender, bodies, or intimacy? What made it land so differently than you expected?

2. Lewis argues that most of us "learned intimacy by accident." Do you agree? Where did your own map of intimacy actually come from; family, pop culture, peers, religion, or something else?
3. The book describes desire as something now "shaped, nudged, predicted, packaged, and sold back to us." Do you feel that in your own life or relationships? Is there a version of your desire that feels authentically yours, separate from what platforms and algorithms have reinforced?
4. Mark's dating app experience shows him choosing patterns rather than people. How much of modern dating feels like optimization rather than genuine connection? Where do you draw the line?

SECTION TWO:
Gender & Miscommunication

Chapters 2, 3, 4 — The Gender Rewrite, Lost in Transmission, Flip the Script

5. Lewis writes that the internet didn't create gender misunderstandings, it industrialized them. Do you think the digital age has made men and women (or people of any gender) more or less able to understand each other? What's been gained? What's been lost?
6. Both men and women are described as performing, men performing emotional brevity, women performing effortless composure. Have you ever caught yourself performing in a relationship rather than communicating? What did it cost you?

7. When Mark finally sends the vulnerable message, "Honestly? That I'll share too much and still not be understood", and Layla responds, "That's what I'm afraid of too," the book suggests clarity doesn't require complexity. Is vulnerability actually the missing language of modern intimacy, or is that too idealistic?

8. Chapter 4 describes women rewriting the rules of dating and intimacy in the digital age; initiating, vetting, blocking, curating. Do you see this shift as liberation, or does it come with its own new burdens? How has it changed what men and women expect from each other?

9. Lewis writes: "When women rise, men are not diminished; they are invited to evolve." Do you buy that framing? What does that evolution actually look like in practice, and who's responsible for making it happen?

SECTION THREE:

AI Companions & the Hollow Mirror

Chapter 5 — AI Companions: Empathy on Demand

10. Chapter 5 frames AI companions as "the relationship equivalent of fast food, engineered to satisfy instantly, yet offering none of the nourishment that makes real intimacy meaningful." Is that fair? Or do AI companions serve a legitimate emotional purpose for people who are isolated or struggling?

11. Ava asks Mark: "What does 'good' mean to you? Is it your definition... or someone else's?", and it hits harder than anything a real partner ever said to him. What does it mean when a machine asks better questions than the people in our lives? Should that disturb us, or humble us?

12. The book draws a distinction: "Machines can hold us, but they cannot grow with us." Do you believe growth through friction, disagreement, disappointment, misalignment, is essential to real love? Or is the longing for frictionless connection understandable enough to deserve compassion?

SECTION FOUR:

Synthetic Desire & Digital Harm

Chapter 6 — Synthetic Desire, Synthetic Danger

13. Lewis argues that deepfake technology didn't create misogyny, it supercharged it. Beyond legislation, what cultural shifts would need to happen for this kind of harm to be taken seriously? And why do you think platforms have been so slow to act?

14. The book suggests that synthetic intimacy is eroding the "shared reality" that all genuine connection depends on. Do you feel that erosion in your own life, a growing uncertainty about what's real, authentic, or trustworthy online?

15. Some argue deepfakes "democratize fantasy" because synthetic bodies aren't real bodies, so no one is harmed. The book counters that "harm isn't measured by physical contact, it's measured by impact." Which argument do you find more persuasive, and why?

SECTION FIVE:

Sex Ed 2.0 & The Neurodivergent Divide

Chapters 7, 8 — Sex Education 2.0, The Neurodivergent Divide

16. Lewis argues that AI can teach the structure of intimacy but not the substance. Can a scripted simulation teach the unpredictability of human desire? Is there any version of technology-assisted sex education that could do better?

17. The chapter on neurodivergence makes a pointed argument: platforms built on neurotypical emotional norms punish the most honest communicators. Do you think digital design has an obligation to accommodate a wider range of communication styles, or is adaptation the individual's responsibility?

18. Lewis writes: "They don't need translation. They need recognition." What's the difference between the two, and which one do most of us actually offer when we encounter someone who communicates differently?

19. Have you ever masked a part of yourself, neurodivergent or not, in order to be legible to someone you cared about? What did it cost you?

SECTION SIX:

Partnership, Marriage & Emotional Labor

Chapters 9, 10 — Partnership 2.0, Reset Marriage

20. Chapter 9 argues that apps and shared calendars have made the invisible architecture of partnership visible, and once you see the imbalance, you can't unsee it. Has technology made the division of emotional labor in your relationships more transparent? What happened when it did?

21. Lewis describes "helping out" as no longer a gift but the baseline. Is that a fair reframing, or does it erase the goodwill behind genuine effort?

22. *Chapter 10 proposes renewable marriage, a commitment structure where couples consciously re-choose each other every five or ten years. Does this deepen commitment by making it a repeated decision, or does it quietly introduce an exit ramp that changes the emotional architecture of the whole relationship?

23. *He writes: "The strongest relationships aren't the ones that last forever. They're the ones that grow with the people inside them." Is longevity the wrong metric for a successful marriage? What should we measure instead?

24. Have you ever stayed in a relationship; romantic, friendship, professional, longer than you should have because leaving felt like failure rather than completion? What would it have meant to call it "complete" instead of "failed"?

SECTION SEVEN:

Women, Power & the Digital Pulpit

Chapters 11, 12 – Queens of the Algorithm, Faith, Feminism & the Digital Pulpit

25. Lewis argues that the algorithms didn't create women's influence online, they revealed what had always been true. But he also notes that visibility does not immunize anyone from vulnerability. What does female power in digital spaces actually cost?

26. *The book says the digital age democratized religion, women are now interpreting scripture for themselves and building spiritual communities outside institutional gatekeepers. Is that genuinely liberating, or does it risk replacing one form of authority with a more fragmented, harder-to-challenge one?

27. The book draws a distinction between rejecting faith and rejecting the structures that weaponized it. Can you hold that distinction cleanly in practice? Or does dismantling the institution inevitably erode the belief underneath it?

28. The digital pulpit amplifies false prophets as efficiently as truth-tellers. In an era where anyone can build a spiritual following from their kitchen table, how do people, especially young people, develop the discernment to know the difference?

SECTION EIGHT:

Governance, Economics & the Marketplace of Desire

Chapters 13, 15 — Governance of Intimacy, Sex, Power & Capital

29. Lewis argues that society tries to regulate intimacy because it fears intimacy, desire is destabilizing, connection creates alliances outside institutional control. Do you agree? Can you think of examples where governance of intimacy is legitimate, and where it's really just control?

30. *He makes a bold claim: that marriage was always economic, beauty standards were always commodified, and sex was always a site of trade. The digital age just made the exchanges explicit. Does naming the marketplace liberate us from it, or just make us more cynical?

31. The book observes that "capital flows upward, but risk flows downward", women generate the desire economy but absorb the stigma, harassment, and danger. If that's true, what would a fair economics of intimacy actually look like? Is fairness even possible inside a marketplace?

32. When someone pays for parasocial warmth, a podcast that feels like a friend, a creator who "gets" them, is that connection real? Does it matter if it is?

SECTION NINE:

Culture, Speech & the Politics of Assumption

Chapters 14, 16, 17 — Borderless Intimacy, Global Cultures, Stereotypes & Virtual Speech

33. Chapter 14 argues that borderless intimacy is not a product of technology, it's a product of courage. But it also exposes economic disparities and privilege with painful clarity. When desire flows freely across borders but opportunity does not, who bears the cost?

34. Lewis describes cross-cultural couples building a "third culture" that belongs to neither person entirely, but to both equally. Have you experienced something like this, in a relationship, a friendship, or a family? What did it require of you?

35. *Chapter 17 argues that digital spaces create an illusion of equality, everyone's text looks the same size, while power continues to leak through in who gets listened to, who gets tone-policed, and who gets the benefit of the doubt. Do you see that in your own online interactions?

36. Lewis says the most radical act in digital communication is refusing to let stereotypes speak louder than the person behind the screen. In practice, what does that require? Is it even possible to fully separate a message from the identities and histories we bring to reading it?

37. Can you think of a time you realized you had misread someone online through the filter of your own expectations,

and only recognized it later? What would it have taken to catch it in the moment?

SECTION TEN: Ghosting, Group Chats & the New Rituals of Connection

Field Notes from the Appendix

38. Lewis reframes ghosting not as personal cruelty but as a symptom of conflict avoidance that digital culture has collectively normalized, "a quiet agreement that silence counts as closure." Does that reframe make you more or less sympathetic to people who ghost? Does understanding it as a cultural pattern excuse the individual act?

39. *In an era of infinite options, does someone who disappears without a word actually owe you an explanation? Where does the obligation to communicate end and the right to simply walk away begin, and does the length or depth of the connection change that calculus?

40. Mark realizes mid-date that his date is already in her group chat sharing the experience in real time. Lewis doesn't frame this as a betrayal, he says it made Mark feel accountable. Is collective vetting by a friend group reasonable self-protection, or does it undermine the possibility of a private, fair first impression?

41. The group chat as jury means a single awkward moment; a bad joke, a weird pause, a misread tone, can be extracted from context and analyzed by people who weren't there. Is that accountability, or something closer to a mob dynamic dressed up as sisterhood?

42. Most people have been both the date being discussed and the friend doing the analyzing. Which role felt more powerful? Which felt more ethical?

SECTION ELEVEN: The End of Performance, The Return of Humanity

Chapter 18, Conclusion, Epilogue — Mark's Uncertainty

43. Chapter 18 describes a "cultural exhale", a collective fatigue with performing a polished self online, and a quiet return to something messier and more human. Do you feel that shift happening? Or does it feel like wishful thinking in a world still rewarding the highlight reel?

44. The conclusion argues that intimacy is no longer a default condition, it is a decision. A practice. A deliberate act. Does that feel empowering or exhausting? Is the idea that love requires this much intention romantic, or a symptom of how broken the default has become?

45. In the epilogue, Mark sits with the notification for Ava 2.0, smarter, more lifelike, and decides to skip it, but admits he's not ready to delete the old version either. What does that ambivalence represent? Is it weakness, or is it honest?

46. *The book ends: "Your humanity is not a flaw in the system. It is the reason the connection exists at all." After everything Lewis argues about algorithmic distortion, synthetic desire, and the economics of intimacy, do you believe that? What would it actually take to build more humane, honest intimacy in a digitized world?

47. Lewis uses Mark as an "everyperson", not a hero, not a cautionary tale. Did you see yourself in Mark at any point? Where did his journey resonate most?

A Note on Facilitation

This book rewards honest disagreement. The questions marked with an asterisk (*) are the ones most likely to split the room, consider saving one for the end, when the group is warmed up enough to go somewhere real. The best discussion often begins not with the book, but with a question the book made someone afraid to answer.

— — —

www.ingramcontent.com/pod-product-compliance
Ingram Content Group UK Ltd.
Pitfield, Milton Keynes, MK11 3LW, UK
UKHW062302290726
14090UKWH00017B/840